BIOMATERIALS - PROPERTIES, PRODUCTION AND DEVICES SERIES

CALCIUM ORTHOPHOSPHATE-BASED BIOCOMPOSITES AND HYBRID BIOMATERIALS

BIOMATERIALS - PROPERTIES, PRODUCTION AND DEVICES SERIES

Biomaterials in Blood-Contacting Devices: Complications and Solutions
Meng-Jiy Wang and Wei-Bor Tsai
2010. ISBN: 978-1-60876-784-7

Surface Modification of Titanium for Biomaterial Applications
Kyo-Han Kim, R. Narayanan and Tapash R. Rautray
2010. ISBN: 978-1-60876-539-3

Calcium Orthophosphate-Based Biocomposites and Hybrid Biomaterials
Sergey V. Dorozhkin
2010. ISBN: 978-1-60876-941-4

Biodegradable Composites for Bone Regeneration
Luigi Calandrelli, Paola Laurienzo and Adriana Oliva
2010. ISBN: 978-1-60876-957-5

BIOMATERIALS - PROPERTIES, PRODUCTION AND DEVICES SERIES

CALCIUM ORTHOPHOSPHATE-BASED BIOCOMPOSITES AND HYBRID BIOMATERIALS

SERGEY V. DOROZHKIN

Nova Science Publishers, Inc.
New York

Copyright © 2010 by Nova Science Publishers, Inc.

All rights reserved. No part of this book may be reproduced, stored in a retrieval system or transmitted in any form or by any means: electronic, electrostatic, magnetic, tape, mechanical photocopying, recording or otherwise without the written permission of the Publisher.

For permission to use material from this book please contact us:
Telephone 631-231-7269; Fax 631-231-8175
Web Site: http://www.novapublishers.com

NOTICE TO THE READER

The Publisher has taken reasonable care in the preparation of this book, but makes no expressed or implied warranty of any kind and assumes no responsibility for any errors or omissions. No liability is assumed for incidental or consequential damages in connection with or arising out of information contained in this book. The Publisher shall not be liable for any special, consequential, or exemplary damages resulting, in whole or in part, from the readers' use of, or reliance upon, this material. Any parts of this book based on government reports are so indicated and copyright is claimed for those parts to the extent applicable to compilations of such works.

Independent verification should be sought for any data, advice or recommendations contained in this book. In addition, no responsibility is assumed by the publisher for any injury and/or damage to persons or property arising from any methods, products, instructions, ideas or otherwise contained in this publication.

This publication is designed to provide accurate and authoritative information with regard to the subject matter covered herein. It is sold with the clear understanding that the Publisher is not engaged in rendering legal or any other professional services. If legal or any other expert assistance is required, the services of a competent person should be sought. FROM A DECLARATION OF PARTICIPANTS JOINTLY ADOPTED BY A COMMITTEE OF THE AMERICAN BAR ASSOCIATION AND A COMMITTEE OF PUBLISHERS.

LIBRARY OF CONGRESS CATALOGING-IN-PUBLICATION DATA
Calcium orthophosphate-based biocomposites and hybrid biomaterials /Author, Sergey V. Dorozhkin.
viii, 169 p. : ill. ; 23 cm.
Includes bibliographical references (p. [75]-152) and index.
ISBN: 978-1-60876-941-4 (softcover)
1. Biomedical materials. 2. Calcium phosphate --Biotechnology. I. Dorozhkin, Sergey V.
R857.M3 D67 2010
610.28/4
2009048902

Published by Nova Science Publishers, Inc.   † New York

CONTENTS

Abstract		vii
Chapter 1	Introduction	1
Chapter 2	General Information on Composites and Biocomposites	7
Chapter 3	The Major Constituent Materials of Biocomposites for Biomedical Applications	11
Chapter 4	Calcium Orthophosphate-Based Biocomposites and Hybrid Biomaterials	19
Chapter 5	Interaction Between the Phases in Calcium Orthophosphate-Based Biocomposites	57
Chapter 6	Bioactivity and Biodegradation of Calcium Orthophosphate- Based Biocomposites	65
Chapter 7	Some Challenges and Critical Issues	69
Chapter 8	Conclusions	73
Chapter 9	References	75
Chapter 10	Nomenclature	153
Index		155

ABSTRACT

In this book, the state-of-the-art of calcium orthophosphate-based biocomposites and hybrid biomaterials suitable for biomedical applications is presented. This subject belongs to a rapidly expanding area of science and research because these types of biomaterials offer many significant and exciting possibilities for hard tissue regeneration. Through the successful combinations of the desired properties of matrix materials with those of fillers (in such systems, calcium orthophosphates might play either role), innovative bone graft biomaterials can be designed. The book starts with an introduction to locate the reader. Further, general information on composites and hybrid materials, including a brief description of their major constituents are presented. Various types of calcium orthophosphate-based bone-analogue biocomposites and hybrid biomaterials those are either already in use or being investigated for various biomedical applications are then extensively discussed. Many different formulations in terms of the material constituents, fabrication technologies, structural and bioactive properties, as well as both *in vitro* and *in vivo* characteristics have been already proposed. Among the others, the nano-structurally controlled biocomposites, those with nano-sized calcium orthophosphates, biomimetically-fabricated formulations with collagen, chitin and/or gelatin, as well as various functionally graded structures seem to be the most promising candidates for clinical applications. The specific advantages of using calcium orthophosphate-based biocomposites and hybrid biomaterials in the selected applications are highlighted. As the way from a laboratory to a hospital is a long one and the prospective biomedical candidates have to meet many different necessities, the critical issues and scientific challenges that require further research and development, have been examined, as well.

Keywords: Calcium orthophosphates, biocomposites, hybrid biomaterials, bone substitutes, functionally graded biomaterials, nano-biocomposites, scaffolds, dental application, hydroxyapatite, polymers, collagen type I, metals, oxides, carbon nanotubes, biocompatible devices, biomaterials, bioceramics, tissue engineering.

Chapter 1

INTRODUCTION

The fracture of bones due to various traumas or natural aging is a typical type of a tissue failure. An operative treatment frequently requires implantation of a temporary or a permanent prosthesis, which still is a challenge for orthopedic surgeons, especially in the cases of large bone defects. A fast aging of the population and serious drawbacks of natural bone grafts make the situation even worse; therefore, there is a high clinical demand for bone substitutes. Unfortunately, a medical application of xenografts (*e.g.*, bovine bone) is generally associated with potential viral infections. In addition, xenografts have a low osteogenicity, an increased immunogenicity and, usually, resorb more rapidly than autogenous bone. Similar limitations are also valid for human allografts (*i.e.*, tissue transplantation between individuals of the same species but of non-identical genetic composition), where the concerns about potential risks of transmitting tumor cells, a variety of bacterial and viral infections, as well as immunological and blood group incompatibility are even stronger [1-3]. Moreover, harvesting and conservation of allografts (exogenous bones) are additional limiting factors. Autografts (endogenous bones) are still the "golden standard" among any substitution materials because they are osteogenic, osteoinductive, osteoconductive, completely biocompatible, non-toxic and do not cause any immunological problems (non-allergic). They contain viable osteogenic cells, bone matrix proteins and support bone growth. Usually, autografts are well accepted by the body and rapidly integrated into the surrounding bone tissues. Due to these reasons, they are used routinely for a long period with good clinical results [3,4]; however, it is fair to say on complication cases, those frequently happened in the past [5,6]. Unfortunately, a limited number of donor sites restrict the quantity of autografts harvested from the iliac crest or other locations of the patient's own body. Also, their medical application is always associated with additional traumas and scars resulting from the extraction of a donor tissue during a superfluous surgical operation, which requires further healing at the donation site and can involve long-term postoperative pain [1,6-9]. Thus, any types of a biologically derived transplants appear to be imperfect solutions, mainly due to a restricted quantity of donor tissues, donor site morbidity, as well as potential risks of an immunological incompatibility and disease transfer [7,9,10]. In this light, manmade materials (alloplastic or synthetic bone grafts) stand out as a reasonable option because they are easily available, might be processed and modified to suit the specific needs of a given application [11,12]. What's more, there are no concerns about potential infections, immunological incompatibility, sterility and donor site morbidity. Therefore, investigations on artificial materials for bone tissue repair appear to be one of the key subjects in the field of biomaterials research for clinical applications [13].

Currently, there are several classes of synthetic bone grafting biomaterials for *in vivo* applications [14-17]. The examples include natural coral, coral-derived materials, bovine porous demineralized bone, human demineralized bone matrix, bioactive glasses, glass-ceramics and calcium orthophosphates [9]. All of these biomaterials are biocompatible and osteoconductive, guiding bone tissue from the edges toward the center of the defect, and aim to provide a scaffold of interconnected pores with pore dimensions ranging from 200 μm [18,19] to 2 mm [20], to facilitate tissue and vessel ingrowths. Among them, porous bioceramics made of calcium orthophosphates appear to be very prominent due to both the excellent biocompatibility and bonding ability to living bone in the body. This is directly related to the fact that the inorganic material of mammalian calcified tissues, *i.e.* of bone and teeth, consists of calcium orthophosphates [21-23]. Due to this reason, other artificial materials are normally encapsulated by fibrous tissue, when implanted in body defects, while calcium orthophosphates are not [24]. Several types of calcium orthophosphate-based bioceramics with different chemical composition are already on the market [9,25]. Unfortunately, as for any ceramic material, calcium orthophosphate bioceramics by itself lack the mechanical and elastic properties of the calcified tissues; namely, scaffolds made of calcium orthophosphates only suffer from a low elasticity, a high brittleness, a poor tensile strength, a low mechanical reliability and fracture toughness, which leads to the concerns about their mechanical performance after implantation

[26-28]. Besides, in many cases, it is difficult to form calcium orthophosphate bioceramics into the desired shapes.

The superior strength and partial elasticity of biological calcified tissues (*e.g.*, bones) are due to the presence of bioorganic polymers (mainly, collagen type I fibers[1]) rather than to a natural ceramic (mainly, a poorly crystalline ion-substituted calcium deficient hydroxyapatite, often referred to as "biological apatite") phase [30,31]. The elastic collagen fibers are aligned in bone along the main stress directions. The biochemical composition of bone is given in Table 1 [32]. A decalcified bone becomes very flexible being easily twisted, whereas a bone without collagen is very brittle; thus, the inorganic nanocrystals of biological apatite provide with the hardness and stiffness, while the bioorganic fibers are responsible for the elasticity and toughness [22,33]. In bones, both types of materials integrate each other into a nanometric scale in such a way that the crystallite size, fibers orientation, short-range order between the components, *etc.* determine its nanostructure and therefore the function and mechanical properties of the entire composite [29,34-38]. From the mechanical point of view, bone is a tough material at low strain rates but fractures more like a brittle material at high strain rates; generally, it is rather weak in tension and shear, particularly along the longitudinal plane. Besides, bone is an anisotropic material because its properties are directionally dependent [21,22,28].

It remains a great challenge to design the ideal bone graft that emulates nature's own structures or functions. Certainly, the successful design requires an appreciation of the structure of bone. According to expectations, the ideal bone graft should be benign, available in a variety of forms and sizes, all with sufficient mechanical properties for use in load-bearing sites, form a chemical bond at the bone/implant interface, as well as be osteogenic, osteoinductive, osteoconductive, biocompatible, completely biodegradable at the expense of bone growth and moldable to fill and restore bone defects [26,36,39]. Further, it should resemble the chemical composition of bones (thus, the presence of calcium orthophosphates is mandatory), exhibit contiguous porosity to encourage invasion by the live host tissue, as well as possess both viscoelastic and semi-brittle behavior, as bones do [40-43].

---

[1] One molecule of collagen type I is a triple helix with 338 repetitions of amino acid residues and is about 300 nm in length [29]. Additionally, bone contains small quantities of other bioorganic materials, such as proteins, polysaccharides and lipids, as well as bone contains cells and blood vessels.

Table 1. The biochemical composition* of bone [32]

Inorganic phases	wt. %	Bioorganic phases	wt. %
calcium orthophosphates (biological apatite)	~ 60	collagen type I	~ 20
water	~ 9	non-collagenous proteins: osteocalcin, osteonectin, osteopontin, thrombospondin, morphogenetic proteins, sialoprotein, serum proteins	~ 3
carbonates	~ 4	other traces: polysaccharides, lipids, cytokines	balance
citrates	~ 0.9	primary bone cells: osteoblasts, osteocytes, osteoclasts	balance
sodium	~ 0.7		
magnesium	~ 0.5		
other traces: Cl^-, F^-, K^+ Sr^{2+}, Pb^{2+}, Zn^{2+}, Cu^{2+}, Fe^{2+}	balance		

*The composition is varied from species to species and from bone to bone.

Moreover, the degradation kinetics of the ideal implant should be adjusted to the healing rate of the human tissue with absence of any chemical or biological irritation and/or toxicity caused by substances, which are released due to corrosion or degradation. Ideally, the combined mechanical strength of the implant and the ingrowing bone should remain constant throughout the regenerative process. Furthermore, the substitution implant material should not disturb significantly the stress environment of the surrounding living tissue [44]. Finally, there is an opinion, that in the case of a serious trauma, bone should fracture rather than the implant [26]. A good sterilizability, storability and processability, as well as a relatively low cost are also of a great importance to permit a clinical application. Unfortunately, no artificial biomaterial is yet available, which embodies all these requirements and unlikely it will appear in the nearest future. To date, most of the available biomaterials appear to be either predominantly osteogenic or osteoinductive or else purely osteoconductive [2].

Careful consideration of the bone type and mechanical properties are needed to design bone substitutes. Indeed, in high load-bearing bones such as the femur, the stiffness of the implant needs to be adequate, not too stiff to result in strain shielding, but rigid enough to present stability. However, in relatively low load-bearing applications such as cranial bone repairs, it is more important to have stability and the correct three-dimensional shapes for aesthetic reasons. One of the most promising alternatives is to apply materials with similar composition and nanostructure to that of bone tissue [36].

Mimicking the structure of calcified tissues and addressing the limitations of the individual materials, development of organic-inorganic hybrid biomaterials provides excellent possibilities for improving the conventional bone implants. In this sense, suitable biocomposites of tailored physical, biological and mechanical properties with the predictable degradation behavior can be prepared combining biologically relevant calcium orthophosphates with bioresorbable polymers [45,46]. As a rule, the general behavior of these bioorganic/calcium orthophosphate composites is dependent on nature, structure and relative contents of the constitutive components, although other parameters such as the preparation conditions also determine the properties of the final materials. Currently, biocomposites with calcium orthophosphates incorporated as either a filler or a coating (or both) either into or onto a biodegradable polymer matrix, in the form of particles or fibers, are increasingly considered for using as bone tissue engineering scaffolds due to their improved physical, biologic and mechanical properties [47-53]. In addition, such biocomposites could fulfill general requirements to the next generation of biomaterials, those should combine the bioactive and bioresorbable properties to activate *in vivo* mechanisms of tissue regeneration, stimulating the body to heal itself and leading to replacement of the implants by the regenerating tissue [46,54,55]. Thus, through the successful combinations of ductile polymer matrixes with hard and bioactive particulate bioceramic fillers, optimal materials can be designed and, ideally, this approach could lead to a superior construction to be used as either implants or posterior dental restorative material [56].

A lint-reinforced plaster was the first composite used in clinical orthopedics as an external immobilizer (bandage) in the treatment of bone fracture by Mathijsen in 1852 [57], followed by Dreesman in 1892 [58]. A great progress in the clinical application of various types of composite materials has been achieved since then. Based on the past experience and newly gained knowledge, various composite materials with tailored mechanical and biological performance can be manufactured and used to meet various clinical requirements [59]. However, this book presents only a brief history and advances in the field of calcium orthophosphate-based biocomposites and hybrid biomaterials suitable for biomedical application. The majority of the reviewed literature is restricted to the recent publications; a limited number of papers published in the XX-th century have been cited. Various aspects of the material constituents, fabrication technologies, structural and bioactive properties, phase interaction have been considered and discussed in details. Finally, several critical issues and scientific challenges that are needed for further advancement are outlined.

Chapter 2

GENERAL INFORMATION ON COMPOSITES AND BIOCOMPOSITES

According to Wikipedia, the free encyclopedia, "*composite materials* (or *composites* for short) are engineered materials made from two or more constituent materials with significantly different physical or chemical properties and which remain separate and distinct on a macroscopic level within the finished structure" [60]. Thus, composites are always heterogeneous. Following the point of view of some predecessors, we also consider that "for the purpose of this book, composites are defined as those having a distinct phase distributed through their bulk, as opposed to modular or coated components" [61, page 1329]. For this reason, with a few important exceptions, the structures obtained by soaking of various materials in supersaturated solutions containing ions of calcium and orthophosphate (*e.g.*, Refs. [62-67]), those obtained by coating of various materials by calcium orthophosphates (*e.g.*, Refs. [68-73]), as well as calcium orthophosphates coated by other compounds [74] have not been considered; however, composite coatings have been considered. Occasionally, porous calcium orthophosphate scaffolds filled by cells inside the pores [75,76], as well as calcium orthophosphates impregnated by biologically active substances [77] are also defined as composites; nevertheless, such structures have not been considered in this book either.

In any composite, there are two major categories of constituent materials: a matrix (or a continuous phase) and (a) dispersed phase(s). To create a composite, at least one portion of each type is required. General information on the major fabrication and processing techniques might be found elsewhere [61]. The continuous phase is responsible for filling the volume, as well as it surrounds and supports the dispersed material(s) by maintaining their relative positions. The dispersed phase(s) is(are) usually responsible for enhancing one or more properties of the matrix. Most of the composites target an enhancement of mechanical properties of the matrix, such as stiffness and strength; however, other properties, such as erosion stability, transport properties (electrical or thermal), radiopacity, density or biocompatibility might also be of a great interest. This synergism produces the properties, which are unavailable from the individual constituent materials [78]. What's more, by controlling the volume fractions and local and global arrangement of the dispersed phase, the properties and design of composites can be varied and tailored to suit the necessary conditions. For example, in the case of ceramics, the dispersed phase serves to impede crack growth. In this case, it acts as reinforcement. A number of methods, including deflecting crack tips, forming bridges across crack faces, absorbing energy during pullout and causing a redistribution of stresses in regions adjacent to crack tips, can be used to accomplish this [79]. Other factors to be considered in composites are the volume fraction of (a) dispersed phase(s), its(their) orientation and homogeneity of the overall composite. For example, higher volume fractions of reinforcement phases tend to improve the mechanical properties of the composites, while continuous and aligned fibers best prevent crack propagation with the added property of anisotropic behavior. Furthermore, the uniform distribution of the dispersed phase is also desirable, as it imparts consistent properties to the composite [60,78].

In general, composites might be simple, complex, graded and hierarchical. The term "a simple composite" is referred to the composites those result from the homogeneous dispersion of one dispersed phase throughout a matrix. The term "a complex composite" is referred to the composites those result from the homogeneous dispersion of several dispersed phases throughout one matrix. The term "a graded composite" is referred to the composites those result from the intentionally structurally inhomogeneous dispersion of one or several dispersed phases throughout one matrix. The term "a hierarchical composite" is referred to the cases, when fine entities of either a simple or a complex composite is somehow aggregated to form coarser ones (*e.g.*, granules or particles) which afterwards are dispersed inside another matrix to produce the second hierarchical scale of the composite structure. Another classification type of the available composites is based on either the matrix materials (metals, ceramics and polymers) or the reinforcement dimensions/shapes (particulates, whiskers/short fibers and continuous fibers) [59].

In most cases, three interdependent factors must be considered in designing of any composite: (i) selection of the suitable matrix and dispersed materials, (ii) choice of appropriate fabrication and processing methods, (iii) internal and external design of the device itself [61]. Besides, any composite must be formed to shape. To do this, the matrix material can be added before or after the dispersed material has been placed into a mold cavity or onto the mold surface. The matrix material experiences a melding event, that, depending upon the nature of the matrix material, can occur in various ways such as chemical polymerization, setting, curing or solidification from a melted state. Due to a general inhomogeneity, the physical properties of many composite materials are not isotropic but rather orthotropic (*i.e.*, there are different properties or strengths in different orthogonal directions) [60,78].

Biocomposites are defined as the composites able to interact well with the human body *in vivo* and, ideally, contain one or more component that stimulates the healing process and uptake of the implant. Thus, for biocomposites the biological compatibility appears to be more important than any other type of compatibility [59]. The most common properties from the bioorganic and inorganic domains to be combined in biocomposites have been summarized in Table 2 [36]. In 1990, Williams summarized the major types of biocomposites that were used in orthopedic applications that time [80]. In 2003, Wang published an excellent update [81]. For general advantages of the modern calcium orthophosphate-based biocomposites over calcium orthophosphate bioceramics and bioresorbable polymers individually, the interested readers are advised to get through "Composite materials strategy" chapter of Ref. [46].

Table 2. General respective properties from the bioorganic and inorganic domains, to be combined in various composites and hybrid materials [36]

Inorganic	Bioorganic
Hardness, brittleness	Elasticity, plasticity
High density	Low density
Thermal stability	Permeability
Hydrophilicity	Hydrophobicity
High refractive index	Selective complexation
Mixed valence slate (red-ox)	Chemical reactivity
Strength	Bioactivity

Chapter 3

THE MAJOR CONSTITUENT MATERIALS OF BIOCOMPOSITES FOR BIOMEDICAL APPLICATIONS

3.1. CALCIUM ORTHOPHOSPHATES

The main driving force behind the use of calcium orthophosphates as bone substitute materials is their chemical similarity to the mineral component of mammalian bones and teeth [21-23]. As a result, in addition to being non-toxic, they are biocompatible, not recognized as foreign materials in the body and, most importantly, both exhibit bioactive behavior and integrate into living tissue by the same processes active in remodeling healthy bone. This leads to an intimate physicochemical bond between the implants and bone, termed osteointegration [81]. More to the point, calcium orthophosphates are also known to support osteoblast adhesion and proliferation [82,83]. Even so, the major limitations to use calcium orthophosphates as load-bearing biomaterials are their mechanical properties; namely, they are brittle with poor fatigue resistance [26-28]. The poor mechanical behavior is even more evident for highly porous ceramics and scaffolds because porosity greater than 100 μm is considered as the requirement for proper vascularization and bone cell colonization [84-86]. That is why, in biomedical applications calcium orthophosphates are used primarily as fillers and coatings [23].

The complete list of known calcium orthophosphates, including their standard abbreviations and the major properties, is given in Table 3, while the detailed information on calcium orthophosphates, their synthesis, structure, chemistry, other properties and biomedical application has been comprehensively reviewed recently [23], where the interested readers are referred to. Even more thorough information might be found in various books and monographs [87-93].

Table 3. Existing calcium orthophosphates and their major properties [23]

Ca/P ionic ratio	Compound	Chemical formula	Solubility at 25 °C, – log(K_s)	Solubility at 25 °C, g/L	pH stability range in aqueous solutions at 25°C
0.5	Monocalcium phosphate monohydrate (MCPM)	$Ca(H_2PO_4)_2 \cdot H_2O$	1.14	~18	0.0–2.0
0.5	Monocalcium phosphate anhydrous (MCPA)	$Ca(H_2PO_4)_2$	1.14	~17	[c]
1.0	Dicalcium phosphate dihydrate (DCPD), mineral brushite	$CaHPO_4 \cdot 2H_2O$	6.59	~0.088	2.0–6.0
1.0	Dicalcium phosphate anhydrous (DCPA), mineral monetite	$CaHPO_4$	6.90	~0.048	[c]
1.33	Octacalcium phosphate (OCP)	$Ca_8(HPO_4)_2(PO_4)_4 \cdot 5H_2O$	96.6	~0.0081	5.5–7.0
1.5	α-Tricalcium phosphate (α-TCP)	$\alpha\text{-}Ca_3(PO_4)_2$	25.5	~0.0025	[a]
1.5	β-Tricalcium phosphate (β-TCP)	$\beta\text{-}Ca_3(PO_4)_2$	28.9	~0.0005	[a]
1.2–2.2	Amorphous calcium phosphate (ACP)	$Ca_xH_y(PO_4)_z \cdot nH_2O$, n = 3–4.5; 15–20% H_2O	[b]	[b]	~5–12 [d]
1.5–1.67	Calcium-deficient hydroxyapatite (CDHA)[e]	$Ca_{10-x}(HPO_4)_x(PO_4)_{6-x}(OH)_{2-x}$ [f] (0<x<1)	~85.1	~0.0094	6.5–9.5
1.67	Hydroxyapatite (HA)	$Ca_{10}(PO_4)_6(OH)_2$	116.8	~0.0003	9.5–12
1.67	Fluorapatite (FA)	$Ca_{10}(PO_4)_6F_2$	120.0	~0.0002	7–12
2.0	Tetracalcium phosphate (TTCP), mineral hilgenstockite	$Ca_4(PO_4)_2O$	38–44	~0.0007	[a]

[a] These compounds cannot be precipitated from aqueous solutions.
[b] Cannot be measured precisely. However, the following values were found: 25.7±0.1 (pH=7.40), 29.9±0.1 (pH=6.00), 32.7±0.1 (pH=5.28).
[c] Stable at temperatures above 100°C.
[d] Always metastable.
[e] Occasionally, CDHA is named as precipitated HA.
[f] In the case x = 1 (the boundary condition with Ca/P = 1.5), the chemical formula of CDHA looks as follows: $Ca_9(HPO_4)(PO_4)_5(OH)$.

3.2. POLYMERS

Polymers are a class of materials consisting of large molecules, often containing many thousands of small units, or monomers, joined together chemically to form one giant chain, thus creating very ductile materials. In this respect, polymers are comparable with major functional components of the biological environment: lipids, proteins and polysaccharides. They differ from each other in chemical composition, molecular weight, polydispersity, crystallinity, hydrophobicity, solubility and thermal transitions. Besides, their properties can be fine-tuned over a wide range by varying the type of polymer, chain length, as well as by copolymerization or blending of two or more polymers [94,95]. Opposite to ceramics, polymers exhibit substantial viscoelastic properties and easily can be fabricated into complex structures, such as sponge-like sheets, gels or complex structures with intricate porous networks and channels [96]. Being X-ray transparent and non-magnetic polymeric materials are fully compatible with the modern diagnostic methods such as computed tomography and magnetic resonance imaging. Unfortunately, most of them are unable to meet the strict demands of the *in vivo* physiological environment. Namely, the main requirements to polymers suitable for biomedical applications are that they must be biocompatible, not eliciting an excessive or chronic inflammatory response upon implantation and, for those that degrade, that they breakdown into non-toxic products only. Unfortunately, polymers, for the most part, lack rigidity, ductility and ultimate mechanical properties required in load bearing applications. Moreover, the sterilization processes (autoclave, ethylene oxide and ^{60}Co irradiation) may affect the polymer properties [97].

There is a variety of biocompatible polymers suitable for biomedical applications. For example, polyacrylates, poly(acrylonitrile-*co*-vinylchloride) and polylysine have been investigated for cell encapsulation and immunoisolation [98,99]. Polyorthoesters and PCL have been investigated as drug delivery devices, the latter for long-term sustained release because of their slow degradation rates [100]. PCL is a hydrolytic polyester having appropriate resorption period and releases nontoxic byproducts upon degradation [101]. Other polyesters and PTFE are used for vascular tissue replacement. Polyurethanes are in use as coatings for pacemaker lead insulation and have been investigated for reconstruction of the meniscus [102,103]. Polymers considered for orthopedic purposes include polyanhydrides, which have also been investigated as delivery devices (due to their rapid and well-defined surface erosion), for bone augmentation or replacement since they can be photopolymerized *in situ* [100,104,105]. To overcome their poor mechanical properties, they have been co-polymerized with imides or formulated to be crosslinkable *in situ* [105]. Other polymers, such as polyphosphazenes, can have their properties (*e.g.*, degradation rate) easily modified by varying the nature of their side groups and have been shown to support osteoblast adhesion, which makes them candidate materials for skeletal tissue regeneration [105]. PPF has emerged as a good bone replacement material, exhibiting good mechanical properties (comparable to trabecular bone), possessing the capability to crosslink *in vivo* through the C=C bond and being hydrolytically degradable. It has also been examined as a material for drug delivery devices [100,104-107]. Polycarbonates have been suggested as suitable materials to make scaffolds for bone replacement and have been modified with tyrosine-derived amino acids to render them biodegradable [100]. Polydioxanone has been also tested for biomedical applications [108]. PMMA is widely used in orthopedics, as a bone cement for implant fixation, as well as to repair certain fractures and bone defects, for example, osteoporotic vertebral bodies [109]. However, PMMA sets by a polymerization of toxic monomers, which also evolves significant amounts of heat that damages tissues. Moreover, it is neither degradable nor bioactive, does not bond chemically to bones and might generate particulate debris leading to an inflammatory foreign body response [104,110]. A number of other non-degradable polymers applied in orthopedic surgery include PE in its different modifications such as low density PE, HDPE and UHMWPE (used as the articular surface of total hip replacement implants [111,112]), polyethylene terepthalate, polypropylene and PTFE which are applied to repair knee ligaments [113]. Polyactive™, a block copolymer of PEG and PBT, was also considered for biomedical application [114-118]. Cellulose [119] and its esters [120] are also popular. Finally yet importantly, polyethylene oxide, PHB and blends thereof have also been tested for biomedical applications [46].

Nonetheless, the most popular synthetic polymers used in medicine are the linear aliphatic poly(α-hydroxyesters) such as PLA, PGA and their copolymers – PLGA (Table 4). These materials have been extensively studied; they appear to be the only synthetic and biodegradable polymers with an extensive FDA approval history [46,105,121-125]. They are biocompatible, mostly non-inflammatory, as well as degrade *in vivo* through hydrolysis and possible enzymatic action into products that are removed from the body by regular metabolic pathways [45,100,105,125-130]. Besides, they might be used for drug-delivery purposes [131]. Poly(α-hydroxyesters) have been investigated as scaffolds for replacement and regeneration of a variety of tissues, cell carriers, controlled delivery devices for drugs or proteins (*e.g.*, growth factors), membranes or films, screws, pins and plates for orthopedic applications [100,105,122,103,125,132-134]. Additionally, the degradation rate of PLGA can be adjusted by varying the amounts of the two component monomers (Table 4), which in orthopedic applications can be exploited to create materials that degrade in concert with bone ingrowth [129,135]. Furthermore, PLGA is known to support osteoblast migration and proliferation [55,105,126,136], which is a necessity for bone tissue regeneration. Unfortunately, such polymers on their own, though they reduce the effect of stress-shielding, are too weak to be used in load bearing situations and are only recommended in certain clinical indications, such as ankle and elbow fractures [125,130]. In addition, they exhibit bulk degradation, leading to both a loss in mechanical properties and lowering of the local solution pH that accelerates further degradation in an autocatalytic manner. As the body is unable to cope with the vast amounts of implant degradation products, this might lead to an inflammatory foreign body response [105,125,132]. Finally, poly(α-hydroxyesters) do not possess the bioactive and osteoconductive properties of calcium orthophosphates [122,137].

Table 4. Major properties of several FDA approved biodegradable polymers [121]

Polymer	Thermal properties[*], °C	Tensile modulus, GPa	Degradation time, months
polyglycolic acid (PGA)	t_g = 35-40 t_m = 225-230	7.06	6-12 (strength loss within 3 weeks)
L-polylactic acid (LPLA)	t_g = 60-65 t_m = 173-178	2.7	>24
D,L-polylactic acid (DLPLA)	t_g = 55-60 amorphous	1.9	12-16
85/15 D,L-polylactic-*co*-glycolic acid (85/15 DLPLGA)	t_g = 50-55 amorphous	2.0	5-6
75/25 D,L-polylactic-*co*-glycolic acid (75/25 DLPLGA)	t_g = 50-55 amorphous	2.0	4-5
65/35 D,L-polylactic-*co*-glycolic acid (65/35 DLPLGA)	t_g = 45-50 Amorphous	2.0	3-4
50/50 D,L-polylactic-*co*-glycolic acid (50/50 DLPLGA)	t_g = 45-50 amorphous	2.0	1-2
poly(ε-caprolactone) (PCL)	t_g = (–60)-(–65) t_m = 58-63	0.4	>24

[*] t_g – glass transition temperature; t_m – melting point.

Several classifications of the biomedically relevant polymers are possible. For example, some authors distinguish between synthetic polymers like PLA and PGA or their copolymers with PCL, and polymers of biological origin like polysaccharides (starch, alginate, chitin/chitosan[1] [138-140], gelatin, cellulose, hyaluronic acid derivatives), proteins (soy, collagen, fibrin [9], silk) and a variety of biofibers, such as lignocellulosic natural fibers [8,141,142]. Other authors differentiate between resorbable or biodegradable (*e.g.*, poly(α-hydroxyesters), polysaccharides and proteins) and non-resorbable (*e.g.*, PE, PMMA and cellulose) polymers [56,142]. As synthetic polymers can be produced under the controlled conditions, they in general exhibit predictable and reproducible mechanical and physical properties such as tensile strength, elastic modulus and degradation rate. Control of impurities is a further advantage of synthetic polymers. The list of synthetic biodegradable polymers used for biomedical application as scaffold materials is available as Table 1 in Ref. [142], while further details on polymers suitable for biomedical applications are available in literature [97,134,143-152] where the interested readers are referred. Good reviews on the synthesis of different biodegradable polymers [153], as well as on the experimental trends in polymer nanocomposites [154] are available elsewhere.

3.3. INORGANIC MATERIALS AND COMPOUNDS (METALS, CERAMICS, GLASS, OXIDES, CARBON, ETC.)

Titanium (Ti) is one of the best biocompatible metals and used most widely as implant [13,155]. Besides, there are other metallic implants made of pure Zr, Hf, V, Nb, Ta, Re [155], Ni, Fe, Cu [156,157,158], Ag, stainless steels and various alloys [158] suitable for biomedical application. Recent studies revealed even a greater biomedical potential of porous metals [159-161]. The metallic implants provide the necessary strength and toughness that are required in load-bearing parts of the body and, due to these advantages, metals will continue to play an important role as orthopedic biomaterials in the future, even though there are concerns with regard to the release of certain ions from and corrosion products of metallic implants. Of course, neither metals

---

[1] Chitosan is a biodegradable and semicrystalline polysaccharide obtained from N-deacetylation of chitin, which is harvested from the exoskeleton of marine crustaceans.

nor alloys are biomimetic[2] in terms of chemical composition because there are no elemental metals in the human body. In addition, even biocompatible metals are bioinert: while not rejected by the human body, any metallic implants cannot actively interact with the surrounding tissues. Nevertheless, in some cases (especially when they are coated by calcium orthophosphates; however, that is another story) the metallic implants can show a reasonable biocompatibility [163]. Only permanent implants are made of metals and alloys, in which degradation or corrosion is not desirable. However, during recent years a number of magnesium alloys have been proposed which are aimed to degrade in the body in order to make room for the ingrowing bone [161,164].

Special types of glasses and glass ceramics are also suitable materials for biomedical applications [165-167] and a special Na_2O–CaO–SiO_2–P_2O_5 glass named Bioglass® [11,24,27,28,168,169] is the most popular among them. They are produced via standard glass production techniques and require pure raw materials. Bioglass® is a biocompatible and osteoconductive biomaterial. It bonds to bone without an intervening fibrous connective tissue interface and, due to these properties, it has been widely used for filling bone defects [170]. The primary shortcoming of Bioglass® is mechanical weakness and low fracture toughness due to an amorphous two-dimensional glass network. The bending strength of most Bioglass® compositions is in the range of 40–60 MPa, which is not suitable for major load-bearing applications. Making porosity in Bioglass®-based scaffolds is beneficial for even better resorption and bioactivity [171].

By heat treatment, a suitable glass can be converted into glass-crystal composites containing crystalline phase(s) of controlled sizes and contents. The resultant glass ceramics can have superior mechanical properties to the parent glass as well as to sintered crystalline ceramics. The bioactive apatite-wollastonite (A-W) glass ceramics is made from the parent glass in the pseudoternary system $3CaO·P_2O_5$–$CaO·SiO_2$–$MgO·CaO·2SiO_2$, which is produced by a conventional melt-quenching method. The bioactivity of A-W glass ceramics is much higher than that of sintered HA. It possesses excellent mechanical properties and has therefore been used clinically for iliac and vertebrae prostheses and as intervertebral spacers [13,172,173].

Metal oxide ceramics, such as alumina (Al_2O_3, high purity, polycrystalline, fine grained) zirconia (ZrO_2) and some other oxides (*e.g.*,

---

[2] The term biomimetic can be defined as a processing technique that either mimics or inspires the biological mechanism, in part or whole [162].

TiO_2), have been widely studied due to their bioinertness, excellent tribological properties, high wear resistance, fracture toughness and strength, as well as a relatively low friction [13,174]. Unfortunately, due to transformation from the tetragonal to the monoclinic phase, a volume change occurs when pure zirconia is cooled down, which causes cracking of the zirconia ceramics. Therefore, additives such as calcia (CaO), magnesia (MgO) and yttria (Y_2O_3) must be mixed with zirconia to stabilize the material in either the tetragonal or the cubic phase. Such material is called PSZ [175-177]. However, the brittle nature of any ceramics has limited their scope of clinical applications and hence more research needs to be conducted to improve their properties.

Chapter 4

CALCIUM ORTHOPHOSPHATE-BASED BIOCOMPOSITES AND HYBRID BIOMATERIALS

Generally, the use of calcium orthophosphate-based biocomposites and hybrid biomaterials for clinical applications has included several (partly overlapping) broad areas:

- biocomposites with polymers,
- cement-based biocomposites and concretes,
- nano-calcium orthophosphate-based biocomposites and nanocomposites,
- biocomposites with collagen,
- biocomposites with other bioorganic compounds and biological macromolecules,
- injectable bone substitutes (IBS),
- biocomposites with glasses, inorganic compounds and metals,
- functionally graded biocomposites,
- biosensors.

The details of each subject are given below.

4.1. BIOCOMPOSITES WITH POLYMERS

Typically, the polymeric components of biocomposites and hybrid biomaterials comprise polymers that both have shown a good biocompatibility and are routinely used in surgical applications. In general, since polymers have a low modulus (2–7 GPa, as the maximum) as compared to that of bone (3–30 GPa), calcium orthophosphate bioceramics need to be loaded at a high weight % ratio. Besides, general knowledge on composite mechanics suggests that any high aspect ratio particles, such as whiskers or fibers, significantly improve the modulus at a lower loading [148]. Thus, some attempts have been already performed to prepare biocomposites containing whisker-like [178-181] or needle-like [182-184] calcium orthophosphates, as well as calcium orthophosphate fibers [45,185].

The history of implantable polymer-calcium orthophosphate biocomposites and hybrid biomaterials started in 1981[1] from the pioneering study by Prof. William Bonfield and colleagues performed on HA/PE composites [187,188]. That initial study introduced a bone-analogue concept, when proposed biocomposites comprised a polymer ductile matrix of PE and a ceramic stiff phase of HA, and was substantially extended and developed in further investigations by that research group [94,189-206]. More recent studies included investigations on the influence of surface topography of HA/PE composites on cell proliferation and attachment [207-213]. The material is composed of a particular combination of HA particles at a volume loading of ~ 40% uniformly dispersed in a HDPE matrix. The idea was to mimic bone by using a polymeric matrix that can develop a considerable anisotropic character through adequate orientation techniques reinforced with a bone-like ceramics that assures both a mechanical reinforcement and a bioactive character of the composite. Following FDA approval in 1994, in 1995 this material has become commercially available under the trade-name HAPEX™ (Smith and Nephew, Richards, USA), and to date remains the only clinically successful bioactive composite that appeared to be a major step in the implant field [28,214]. The major production stages of HAPEX™ include blending, compounding and centrifugal milling. A bulk material or device is then created from this powder by compression and injection molding [59]. Besides, HA/HDPE biocomposites might be prepared by a hot rolling technique that facilitated uniform dispersion and blending of the reinforcements in the matrix [215].

A mechanical interlock between the two phases of HAPEX™ is formed by shrinkage of HDPE onto the HA particles during cooling [94,216]. Both HA particle size and their distribution in the HDPE matrix were recognized as

---

[1] However, a more general topic "ceramic-plastic material as a bone substitute" is, at least, 18 years older [186].

important parameters affecting the mechanical behavior of HAPEX™ [198]. Namely, smaller HA particles were found to lead to stiffer composites due to general increasing of interfaces between the polymer and the ceramics; furthermore, rigidity of HAPEX™ was found to be proportional to HA volume fraction [190]. In this formulation, HA could be replaced by other calcium orthophosphates [217].

Initial clinical applications of HAPEX™ came in orbital reconstruction [218] but since 1995, the main uses of this composite have been in the shafts of middle ear implants for the treatment of conductive hearing loss [219,220]. In both applications, HAPEX™ offers the advantage of *in situ* shaping, so a surgeon can make final alterations to optimize the fit of the prosthesis to the bone of a patient and subsequent activity requires only limited mechanical loading with virtually no risk of failure from insufficient tensile strength [94,168]. As compared to cortical bones, HA/PE composites have a superior fracture toughness for HA concentrations below 40% and similar fracture toughness in the 45–50% range. Their Young's modulus is in the range of 1–8 GPa, which is quite close to that of bone. The examination of the fracture surfaces revealed that only mechanical bond occurs between HA and PE. Unfortunately, the HA/PE composites are not biodegradable, the available surface area of HA is low and the presence of bioinert PE decreases the ability to bond to bones. Furthermore, HAPEX™ has been designed with a maximized density to increase its strength but the resulting lack of porosity limits the ingrowth of osteoblasts when the implant is placed into the body [26,169]. Further details on HAPEX™ are available elsewhere [94]. Except of HAPEX™, other types of HA/PE biocomposites are also known [221-225].

Both linear and branched PE was used as a matrix and the biocomposites with the former were found to give a higher modulus [222]. The reinforcing mechanisms in calcium orthophosphate/polymer biocomposites have yet to be convincingly disclosed. Generally, if a poor filler choice is made, the polymeric matrix might be affected by the filler through reduction of molecular weight during composite processing, formation of an immobilized shell of polymer around the particles (transcrystallization, surface-induced crystallization or epitaxial growth) and changes in conformation of the polymer due to particle surfaces and inter-particle spacing [94]. On the other hand, the reinforcing effect of calcium orthophosphate particles might depend on the molding technique employed: a higher orientation of the polymeric matrix was found to result in a higher mechanical performance of the composite [226,227].

Many other blends of calcium orthophosphates with various polymers are possible, including rather unusual formulations with dendrimers [228]. The list of the appropriate calcium orthophosphates is shown in Table 3 (except of MCPM and MCPA – both are too acidic and, therefore, are not biocompatible [23]), while many biomedically suitable polymers have been listed above. The combination of calcium orthophosphates and polymers into biocomposites has a twofold purpose. The desirable mechanical properties of polymers compensate for a poor mechanical behavior of calcium orthophosphate bioceramics, while in turn the desirable bioactive properties of calcium orthophosphates improve those of polymers, expanding the possible uses of each material within the body [127-129,229-232]. Namely, polymers have been added to calcium orthophosphates in order to improve their mechanical strength [127,229] and calcium orthophosphate fillers have been blended with polymers to improve their compressive strength and modulus, in addition to increasing their osteoconductive properties [48,129,137,233-237]. Furthermore, biocompatibility of such biocomposites is enhanced because calcium orthophosphate fillers induce an increased initial flash spread of serum proteins compared to the more hydrophobic polymer surfaces [238]. What's more, experimental results of these biocomposites indicate favorable cell-material interactions with increased cell activities as compared to each polymer alone [231]. As a rule, with increasing of calcium orthophosphate content, both Young's modulus and bioactivity of the biocomposites increase, while the ductility decreases [26,233]. Furthermore, such formulations can provide a sustained release of calcium and orthophosphate ions into the milieus, which is important for mineralized tissue regeneration [230]. Indeed, a combination of two different materials draws on the advantages of each one to create a superior biocomposite with respect to the materials on their own.

It is logical to assume that the proper biocomposite of a calcium orthophosphate (for instance, CDHA) with a bioorganic polymer (for instance, collagen) would yield the physical, chemical and mechanical properties similar to those of human bones. Different ways have been already realized to bring these two components together into composites, like mechanical blending, ball milling, dispersion of ceramic fillers into a polymer-solvent solution, a melt extrusion of a ceramic/polymer powder mixture, co-precipitation and electrochemical co-deposition [32,59,239-241]. Besides, there is *in situ* formation, which involves either synthesizing the reinforcement inside a preformed matrix material or synthesizing the matrix material around the reinforcement [59,242]. For example, several papers have reported this method to produce various composites of apatites with carbon nanotubes [243-248].

Another example comprises using amino acid-capped gold nanoparticles as scaffolds to grow CDHA [249]. In certain cases, a mechano-chemical route [250], emulsions [251-254], freeze-drying [255] and freeze-thawing techniques [256], flame-sprayed technique [257] or gel-templated mineralization [258] might be applied to produce calcium othophosphates-based biocomposites. Various fabrication procedures are available elsewhere [32,59,239], where the interested readers are referred.

The interfacial bonding between a calcium orthophosphate and a polymer is an important issue of any biocomposite. If adhesion between the phases is poor, the mechanical properties of a biocomposite suffer. To solve the problem, various approaches have been already introduced. For example, a diisocyanate coupling agent was used to bind PEG/PBT (Polyactive^{TM}) block copolymers to HA filler particles. Using surface-modified HA particles as a filler in a PEG/PBT matrix significantly improved the elastic modulus and strength of the polymer as compared to the polymers filled with ungrafted HA [235,259]. Another group used processing conditions to achieve a better adhesion of the filler to the matrix. Ignjatovic et al. prepared PLLA/HA composites by pressing blends of varying PLLA and HA content at different temperatures and pressures [127,128,260]. They found that maximum compressive strength was achieved at ~ 15 wt.% of PLLA. By using blends with 20 wt.% of PLLA, the authors also established that increasing the pressing temperature and pressure improved the mechanical properties. The former was explained by decrease in viscosity of the PLLA associated with a temperature increase, hence leading to improved wettability of HA particles. The latter was explained by increased compaction and penetration of pores at higher pressure, in conjunction with a greater fluidity of the polymer at higher temperatures. The combination of high pressures and temperatures was found to decrease porosity and guarantee a close apposition of a polymer to the particles, thereby improving the compressive strength [229] and fracture energy [261] of the biocomposites. The PLLA/HA biocomposites scaffolds were found to improve cell survival over plain PLLA scaffolds [262].

It is also possible to introduce porosity into calcium orthophosphate-based biocomposites, which is advantageous for most applications as bone substitution material. The porosity facilitates the migration of osteoblasts from surrounding bones to the implant site [129,263,264]. Various material processing strategies to prepare composite scaffolds with interconnected porosity comprise thermally induced phase separation, solvent casting and particle leaching, solid freeform fabrication techniques, microsphere sintering and coating [142,265-267]. A supercritical gas foaming technique might be used as well [239,268,269].

Apatite-Based Biocomposites

A biological apatite is known to be the major inorganic phase of mammalian calcified tissues [21,22]. Consequently, CDHA, HA, carbonateapatite (both with and without dopants) and, occasionally, FA have been applied to prepare biocomposites with other compounds, usually with the aim to improve the bioactivity. For example, PS composed with HA can be used as a starting material for long-term implants [270-272]. Retrieved *in vivo*, HA/PS biocomposite coated samples from rabbit distal femurs demonstrated direct bone apposition to the coatings, as compared to the fibrous encapsulation that occurred when uncoated samples were used [270]. The resorption time of such biocomposites is a very important factor, which depends on polymer's microstructure and the presence of modifying phases [271].

Various apatite-containing biocomposites with PVA [256,273-279], PVAP [282] and several other polymeric components [280,281,283-293] have been already developed. Namely, PVA/CDHA biocomposite blocks were prepared by precipitation of CDHA in aqueous solutions of PVA [256]. An artificial cornea consisted of a porous nano-HA/PVA hydrogel skirt and a transparent center of PVA hydrogel has been prepared as well. The results displayed a good biocompatibility and interlocking between artificial cornea and host tissues [277,278]. PVAP has been chosen as a polymer matrix, because its phosphate groups can act as a coupling/anchoring agent, which has a higher affinity toward the HA surface [282]. Greish and Brown developed HA/Ca poly(vinyl phosphonate) biocomposites [284-286]. A template-driven nucleation and mineral growth process for the high-affinity integration of CDHA with PHEMA hydrogel scaffold has been developed as well [293].

PEEK [178,180,294-300] and HIPS [301] were applied to create biocomposites with HA having a potential for clinical use in load bearing applications. The study on reinforcing PEEK with thermally sprayed HA particles revealed that the mechanical properties increased monotonically with the reinforcement concentration, with a maximum value in the study of 40% volume fraction of HA particles [296-298]. The reported ranges of stiffness within 2.8–16.0 GPa and strength within 45.5–69 MPa exceeded the lower values for human bone (7–30 GPa and 50–150 MPa, respectively) [297].

Modeling of the mechanical behavior of HA/PEEK biocomposites is available elsewhere [299].

Biodegradable poly(α-hydroxyesters) are well established in clinical medicine. Currently, they provide with a good choice when a suitable polymeric filler material is sought. For example, HA/PLGA composites were developed which appeared to possess a cellular-compatibility suitable for bone tissue regeneration [302-309]. Zhang and Ma seeded highly porous PLLA foams with HA particles in order to improve the osteoconductivity of polymer scaffolds for bone tissue engineering [48,234]. They pointed out that hydration of the foams prior to incubation in simulated body fluid increased the amount of carbonated CDHA material due to an increase of COOH and OH groups on the polymer surface, which apparently acted as nucleation sites for apatite. The following values of Young's modulus, compressive, bending and tensile strengths for PLLA/HA composites have been achieved: 5–12 GPa, 78–137 MPa, 44–280 MPa and 10–30 MPa, respectively [310]. However, these data does not appear to be in a good agreement with HA/PLLA biocomposite unit cell model predictions [311].

On their own, PGA and PLA are known to degrade to acidic products (glycolic and lactic acids, respectively) that both catalyze polymer degradation and cause inflammatory reactions of the surrounding tissues [312]. Thus, in biocomposites of poly(α-hydroxyesters) with calcium orthophosphates, the presence of slightly basic compounds (HA, TTCP) to some extent neutralizes the acid molecules, provides with a weak pH-buffering effect at the polymer surface and, therefore, more or less compensates these drawbacks [137,313-315]. However, additives of even more basic chemicals (*e.g.*, CaO, $CaCO_3$) might be necessary [142,314,316,317]. Extensive cell culture experiments on pH-stabilized composites of PGA and carbonateapatite were reported, which afterwards were supported by extensive *in vitro* pH-studies [318]. A consequent development of this approach has led to designing of functionally graded composite skull implants consisting of polylactides, carbonateapatite and $CaCO_3$ [319,320]. Besides the pH-buffering effect, inclusion of calcium orthophosphates was found to modify both surface and bulk properties of the biodegradable poly(α-hydroxyesters) by increasing the hydrophilicity and water absorption of the polymer matrix, thus altering the scaffold degradation kinetics. For example, polymer biocomposites filled with HA particles was found to hydrolyze homogeneously due to water penetrating into interfacial regions [321].

Biocomposites of poly(α-hydroxyesters) with calcium orthophosphates are mainly prepared by incorporating the inorganic phase into a polymeric solution, followed by drying under vacuum. The resulting solid composites might be shaped using different processing techniques. One can also prepare these biocomposites by mixing HA particles with L-lactide prior the polymerization [313] or by a combination of slip-casting technique and hot pressing [322]. A surfactant might be useful to keep the suspension homogeneity [323]. Besides, HA/PLA [252,253] and HA/PLGA [254] microspheres might be prepared by a microemulsion technique. More complex carbonated-FA/PLA porous biocomposite scaffolds are also known [324]. An interesting list of references, assigned to the different ways of preparing HA/poly(α-hydroxyesters) biodegradable composites, might be found in publications by Durucan and Brown [49,325,326]. The authors prepared CDHA/PLA and CDHA/PLGA composites by solvent casting technique with a subsequent hydrolysis of α-TCP to CDHA in aqueous solutions. The presence of both polymers was found to inhibit α-TCP hydrolysis, if compared with that of single-phase α-TCP; what is more, the inhibiting effect of PLA exceeded that of PLGA [49,325,326]. The physical interactions between calcium orthophosphates and poly(α-hydroxyesters) might be easily seen in Fig. 1 [49]. Nevertheless, it should not be forgotten that typically non-melt based routes lead to development of composites with lower mechanical performance and many times require the use of toxic solvents and intensive hand labor [147].

Figure 1. SEM micrographs of a) α-TCP compact; b) α-TCP-PLGA biocomposite (bars = 5 μm). Reprinted from Ref. [49] with permission.

The mechanical properties of poly(α-hydroxyesters) could be substantially improved by addition of calcium orthophosphates [327,328]. Shikinami and Okuno developed CDHA/PLLA composites of very high mechanical properties [137]; mini-screws and mini-plates made of these composites have been manufactured and tested [321]. They have shown easy handling and shaping according to the implant site geometry, total resorbability, good ability to bond directly to the bone tissue without interposed fibrous tissue, osteoconductivity, biocompatibility and high stiffness retainable for the period necessary to achieve bone union [321]. The initial bending strength of 280 MPa exceeded that of cortical bone (120–210 MPa), while the modulus was as high as 12 GPa [137]. The strength could be maintained above 200 MPa up to 25 weeks in phosphate-buffered saline solution. Such biocomposites were obtained from precipitation of a PLLA/dichloromethane solution, where small granules of uniformly distributed CDHA microparticles (average size of 3 μm) could be prepared [136]. Porous scaffolds of PDLLA and HA have been manufactured as well [269,329,330]. Upon implantation into rabbit femora, a newly formed bone was observed and biodegradation was significantly enhanced if compared to single-phase HA bioceramics. This might be due to a local release of lactic acid, which in turn dissolves HA. In other studies, PLA and PGA fibers were combined with porous HA scaffolds. Such reinforcement did not hinder bone ingrowth into the implants, which supported further development of such biocomposites as bone graft substitutes [47,48,310,331,332].

Recently, blends (named as SEVA-C) of EVOH with starch filled with 10–30 wt.% HA have been fabricated to yield biocomposites with modulus up to ~ 7 GPa with a 30% HA loading [333-338]. The incorporation of bioactive fillers such as HA in SEVA-C aimed to assure the bioactive behavior of the composite and to provide the necessary stiffness within the typical range of human cortical bone properties. These biocomposites exhibited a strong *in vitro* bioactivity that was supported by the polymer's water-uptake capability [339]. However, the reinforcement of SEVA-C by HA particles was found to affect the rheological behavior of the blend. A degradation model of these biocomposites is available [340].

Higher homologues poly(3-hydroxybutyrate), 3-PHB, and poly(3-hydroxyvalerate), 3-PHV, show almost no biodegradation. Nevertheless, biocomposites of these polymers with calcium orthophosphates showed a good biocompatibility both *in vitro* and *in vivo* [94,341-346]. Both bioactivity and mechanical properties of these biocomposites can be tailored by varying the volume percentage of calcium orthophosphates. Similarly, biocomposites of

PHBHV with both HA and amorphous carbonated apatite (almost ACP) appeared to have a promising potential for repair and replacement of damaged bones [347-350].

Along this line, PCL is used as a slowly biodegradable but well biocompatible polymer. PCL/HA composites have been already discussed as suitable materials for substitution, regeneration and repair of bone tissues [265,351-358]. For example, biocomposites were obtained by infiltration of ε-caprolactone monomer into porous apatite blocks and *in situ* polymerization [354]. The composites were found to be biodegradable and might be applied as cancellous or trabecular bone replacement material or for cartilage regeneration. Both the mechanical performance and biocompatibility in osteoblast cell culture of PCL were shown to be strongly increased when HA was added [359]. Several preparation techniques of PCL/HA composites are known. For example, to make composite fibers of PCL/nano-HA, the desired amount of nano-HA powder was dispersed in a solvent using magnetic stirrer followed by ultrasonication for 30 min. Then, PCL was dissolved in this suspension, followed by the solvent evaporation [360]. The opposite preparation order is also possible: PCL was initially dissolved in chloroform at room temperature (7–10% weight/volume), then HA (~ 10 μm particle size) was suspended in the solution, sonicated for 60 s, followed by the solvent evaporation [129] or salt-leaching [361]. The mechanical properties obtained by this technique were about one-third that of trabecular bone. In a comparative study, PCL and biological apatite were mixed in the ratio 19:1 in an extruder [362]. At the end of the preparation, the mixture was cooled in an atmosphere of nitrogen. The authors observed that the presence of biological apatite improved the modulus while concurrently increasing the hydrophilicity of the polymeric substrate. Besides, an increase in apatite concentration was found to increase both the modulus and yield stress of the composite, which indicated to good interfacial interactions between the biological apatite and PCL. It was also observed that the presence of biological apatite stimulated osteoblasts attachment to the biomaterial and cell proliferation [362]. In another study, a PCL/HA biocomposite was prepared by blending in melt form at 120°C until the torque reached equilibrium in the rheometer that was attached to the blender [363]. Then the sample was compression molded and cut into specimens of appropriate size for testing. It was observed that the composite containing 20 wt.% HA had the highest strength [363]. However, a direct grafting of PCL on the surface of HA particles seems to be the most interesting preparation technique [351]. HA porous scaffolds were coated by a PCL/HA composite coating [50]. In this system, PCL, as a coating component, was able to improve the brittleness and low strength of the HA scaffolds, while the particles in the coating were to improve the osteoconductivity and bioactivity of the coating layer. More complex PDLLA/PCL/HA biocomposites have been prepared as well [364]. Further details on both PCL/HA biocomposites and processing methodologies thereof might be found elsewhere [265].

The spread of attached human osteoblasts onto PLA and PCL films reinforced with CDHA and sintered HA was shown to be higher than for the polymers alone [153]. Moreover, biochemical assays relating cell activity to DNA content allowed concluding that cell activity was more intense for the composite films [153]. Kim *et al.* coated porous HA blocks with PCL from dichloromethane solution and performed drug release studies. The antibiotic tetracycline hydrochloride was added into this layer, yielding a bioactive implant with drug release for longer than a week [50].

Yoon *et al.* investigated the highest mechanical and chemical stability of FA by preparing FA/collagen biocomposites and studied their effect in osteoblast-like cell culture [365]. The researchers found an increased cellular activity in FA composites compared to HA composites. This finding was confirmed in another study by means of variations in the fluoride content for FA-HA/PCL composites [366]. An interesting phenomenon of fractal growth of FA/gelatin composite crystals (Fig. 2) was achieved by diffusion of calcium- and orthophosphate+fluoride-solutions from the opposite sides into a tube filled with a gelatin gel [367-375]. The reasons of this phenomenon are not quite clear yet; besides, up to now nothing has yet been reported on a possible biomedical application of such very unusual structural composites.

TCP-Based Biocomposites

Both α-TCP and β-TCP have a higher solubility than HA (Table 3). Besides, they are faster resorbed *in vivo*.[2] Therefore, these calcium orthophosphates were used instead of HA to prepare completely biodegradable biocomposites [377-395]. For example, a biodegradable and osteoconductive biocomposite made of β-TCP particles and gelatin was proposed [386]. This material was tested *in vivo* with good results. It was found to be biocompatible, osteoconductive and biodegradable with no need for a second

---

[2] However, there are some reports about a lack of TCP biodegradation after implantation in calvarial defects [376].

surgical operation to remove the device after healing occurred. Herbal extracts might be added to this biocomposite [387]. Another research group prepared biocomposites of crosslinked gelatin with β-TCP; they found both a good biocompatibility and bone formation upon subcutaneous implantation in rats [388]. Yang *et al.* [393] extended this to porous (porosity about 75%) β-TCP/gelatin biocomposites those also contained BMP-4. Besides, cell-compatible and possessive some osteoinductive properties porous β-TCP/alginate-gelatin hybrid scaffolds were prepared and successfully tested *in vitro* [390]. More to the point, biocomposites of β-TCP with PLLA [383,384] and PLGC [385] were prepared. Although β-TCP was able to counter the acidic degradation of the polyester to some extent, it did not prevent a pH drop down to ~ 6. Nevertheless, implantation of this biocomposite in beagles' mandibular bones was successful [385].

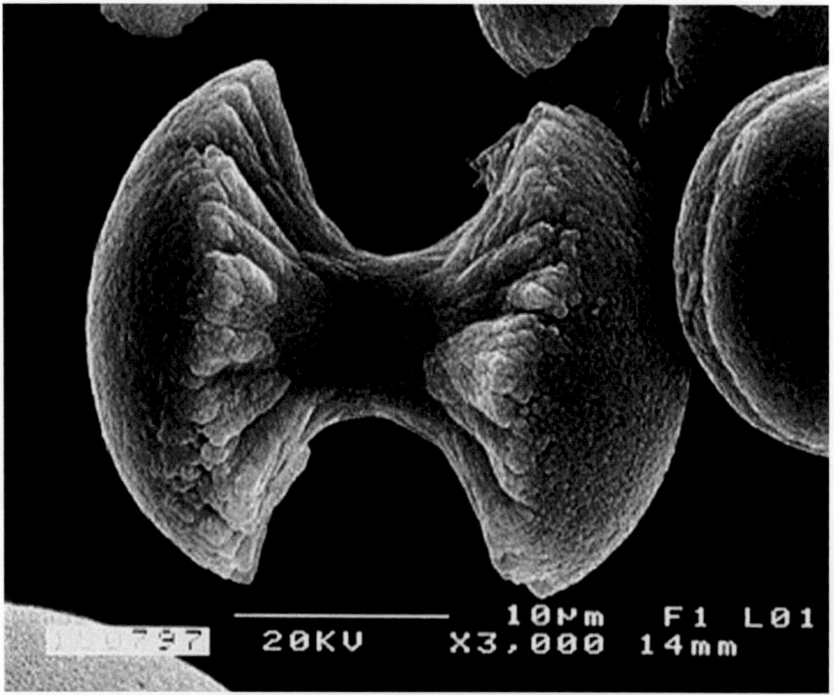

Figure 2. A biomimetically grown aggregate of FA that was crystallized in a gelatin matrix. Its shape can be explained and simulated by a fractal growth mechanism. Scale bar: 10 μm. Reprinted from Ref. [367] with permission.

Based on the self-reinforcement concept, biocomposites of TCP with polylactides were prepared and studied using conventional mechanical testing [396]. Bioresorbable scaffolds were fabricated from such biocomposites [397]. Chitosan was also used as the matrix for the incorporation of β-TCP by a solid/liquid phase separation of the polymer solution and subsequent sublimation of the solvent. Due to complexation of the functional groups of chitosan with calcium ions of β-TCP, these biocomposites had a better compressive modulus and strength [398]. PCL/β-TCP biocomposites were developed as well [399-402] and their *in vitro* degradation behavior was systematically monitored by immersion in simulated body fluid at 37 °C [401]. To extend this topic further, the PCL/β-TCP biocomposites might be loaded by drugs [402].

Cell culture tests on β-TCP/PLLA biocomposites were reported; the biocomposites showed no cytotoxicity and evidenced good cell attachment to its surface [377]. An *in vitro* study with primary rat calvarial osteoblasts showed an increased cellular activity in the BMP-loaded samples [393]. Other researchers investigated BMP-2-loaded porous β-TCP/gelatin biocomposites (porosity 95%, average pore size 180-200μm) [403] and confirmed the precious study. Biocomposites of β-TCP and glutaraldehyde cross-linked gelatin were manufactured and tested *in vitro* to measure the material cytotoxicity [389]. The experimental results revealed that the amount of glutaraldehyde cross-linking agent should be less than 8% to decrease the toxicity on the osteoblasts and to avoid inhibition of cellular growth caused by the release of residual or uncross-linked glutaraldehyde.

A long-term implantation study of PDLLA/α-TCP composites in a loaded sheep implant model showed good results after 12 months but a strong osteolytic reaction after 24 months. This was ascribed to the almost complete dissolution of α-TCP to this time and an adverse reaction of the remaining PDLLA [404].

More complex calcium orthophosphate-based biocomposites are known as well. For example, there is a composite consisting of three interpenetrating networks: TCP, CDHA and PLGA [405]. Firstly, a porous TCP network was produced by coating a polyurethane foam by hydrolysable α-TCP slurry. Then, a CDHA network was derived from a calcium orthophosphate cement filled in the porous TCP network. Finally, the remaining open pore network in the CDHA/α-TCP structures was infiltrated with PLGA. This biocomposite consists of three phases with different degradation behavior. It was postulated that bone would grow on the fastest degrading network of PLGA, while the remaining calcium orthophosphate phases would remain intact thus maintaining their geometry and load bearing capability [405].

Other Calcium Orthophosphate-Based Biocomposites

The number of research papers devoted to biocomposites based on other calcium orthophosphates is substantially lesser than those devoted to apatites and TCP. Biphasic calcium phosphate (BCP)[3] appears to be most popular among the remaining calcium orthophosphates. Collagen coated BCP ceramics was studied and the biocompatibility towards osteoblasts was found to increase upon coating with collagen [406]. Another research group created porous PDLLA/BCP scaffolds and coated them with a hydrophilic PEG/vancomycin composite for both drug delivery purposes and surface modification [407]. More to the point, PLGA/BCP composites were fabricated [408,409] and their cytotoxicity and fibroblast properties were found to be acceptable for natural bone tissue reparation, filling and augmentation [410,411]. PCL/BCP biocomposites are known as well [412].

A choice of DCPD-based biocomposites of DCPD, albumin and duplex DNA was prepared by water/oil/water interfacial reaction method [251]. Core-shell type DCPD/chitosan biocomposite fibers were prepared by a wet spinning method in another study [413]. The energy-dispersive X-ray spectroscopy analysis indicated that Ca and P atoms were mainly distributed on the outer layer of the composite fibers; however, a little amount of P atoms remained inside the fibers. This indicated that the composite fibers formed a unique core-shell structure with shell of calcium orthophosphate and core of chitosan [413]. Although, this is not to the point, it is interesting to mention that some DCPD/polymer composites could be used as proton conductors in battery devices [414,415]. Nothing has been reported on their biocompatibility but, perhaps, sometime the improved formulations will be used to fabricate biocompatible batteries for implantable electronic devices.

Various ACP-based biocomposites for dental applications were developed [416-419]. Besides, several ACP-based formulations were investigated as potential biocomposites for bone grafting [350,420-422]. Namely, ACP/PPF biocomposites were prepared by *in situ* precipitation [421], while PHB/carbonated ACP and PHBHV/carbonated ACP biocomposites appeared

---

[3] BCP is a solid composite of HA and β-TCP; however, similar composites of HA and α-TCP are possible as well [23].

to be well suited as slowly biodegradable bone substitution material [350]. Another example comprises hybrid nano-capsules of ~ 50-70 nm in diameter which were fabricated by ACP mineralization of shell cross-linked polymer micelles and nanocages [422]. These nano-capsules consisted of a continuous ultrathin inorganic surface layer that infiltrated the outer cross-linked polymeric domains. They might be used as structurally robust, pH-responsive biocompatible hybrid nanostructures for drug delivery, bioimaging and therapeutic applications [422].

4.2. CALCIUM ORTHOPHOSPHATE CEMENT-BASED BIOCOMPOSITES AND CONCRETES

Inorganic self-setting calcium orthophosphate cements, which harden in the body, were introduced by LeGeros *et al.* [423] and Brown and Chow [424,425] in the early 1980-s[4]. Since then, these cements have been broadly studied and many formulations have been proposed [428]. The cements set and harden due to various chemical interactions among calcium orthophosphates that finally lead to formation of a monolithic body consisting of either CHDA or DCPD with possible admixtures of other phases. Unfortunately, having the ceramic nature, calcium orthophosphate cements are brittle after hardening and the setting time is sometimes unsuitable for clinical procedures [428]. Therefore, various attempts have been performed to transform the cements into biocomposites *e.g.*, by adding hydroxylcarboxylic acids, to control the setting time [429], gelatin to improve both the mechanical properties and the setting time [392,430-432] or osteocalcin/collagen to increase the bioactivity [433]. More to the point, various reinforcement additives of different shapes and nature are widely used to improve the mechanical properties of calcium orthophosphate cements [428]. Even carbon nanotubes were used for this purpose [434]! Although the biomaterials community does not use this term, a substantial amount of the reinforced cement formulations might be defined as calcium orthophosphate-based concretes[5]. The idea behind the concretes is

---

[4] There is an opinion [426] that the self-setting calcium orthophosphate cements for orthopedic and dental restorative applications have first been described in the early 1970-s by Driskell *et al.* in US Patent No. 3913229 [427].

[5] According to Wikipedia, the free encyclopedia: "*Concrete* is a construction material that consists of a cement (commonly Portland cement), aggregates (generally gravel and sand) and water. It solidifies and hardens after mixing and placement due to a chemical process

simple: if a strong filler is present in the matrix, it might stop crack propagation.

Various apatite-containing biocomposite formulations based on PMMA [436-446] and PEMA [94,447,448] have been already developed. Such biocomposites might be prepared by dispersion of apatite powder into a PMMA viscous fluid [449] and used for drug-delivery purposes [450]. When the mechanical properties of the biocomposite concretes composed of PMMA matrix and HA particles of various sizes were tested, the tensile results showed that strength was independent on particle sizes. In addition, up to 40% by weight HA could be added without impairing the mechanical properties [439,440]. After immersion into Ringer's solution, the tensile strength was not altered whereas the fatigue properties were significantly reduced. The biocompatibility of PMMA/HA biocomposites was tested *in vivo* and enhanced osteogenic properties of the implants compared to single-phase PMMA were observed [437,441-444]. It was shown that not only the mechanical properties of PMMA were improved but the osteoblast response of PMMA was also enhanced with addition of HA [441]. Thereby, by adding of calcium orthophosphates, a non-biodegradable PMMA was made more bioactive and osteoconductive, yielding a well-processable biocomposite concrete. As a drawback, the PMMA/HA formulations possess a low flexural, compressive and tensile strength.

A biocomposite made from HA granules and bis-phenol-α-glycidylmethacrylate-based resin appeared to possess comparable mechanical and biological properties to typical PMMA cement, leading to potential uses for implant fixation [451]. To improve the mechanical properties of calcium orthophosphate cements and stabilize them at the implant site, various researchers have resorted to formulations that set *in situ*, primarily through crosslinking reactions of the polymeric matrix. For example, TTCP was reacted with PAA, forming a crosslinked CDHA/calcium polyacrylate biocomposite [452]. In aqueous solutions, TTCP hydrolyzes to CDHA [23] and the liberated calcium cations react with PAA, forming the crosslinked network [452]. Reed *et al.*, synthesized a dicarboxy polyphosphazene that can be crosslinked by calcium cations and cement-based (TTCP+DCPD) CDHA/polyphosphazene biocomposites with a compressive strength ~10 MPa and of ~65% porosity were prepared as a result [453]. To mimic PMMA cements, PFF/β-TCP biocomposites were prepared with addition of vinyl

---

known as hydration. The water reacts with the cement, which bonds the other components together, eventually creating a stone-like material" [435].

monomer to crosslink PPF. As a result, quick setting and degradable biocomposite cements with a low heat output and compressive strengths in the range of 1-12 MPa were prepared by varying the molecular weight of PPF, as well as the contents of the monomer, β-TCP, initiator and porogen (NaCl) [454,455]. An acrylic cement with Sr-containing HA as a filler [110] and an injectable polydimethylsiloxane/HA cement [456] have been prepared as well.

In order to improve the mechanical properties of calcium orthophosphate cements, numerous researchers blended various polymers with the cements. For example, gelatin might be added to calcium orthophosphate cement formulations, primarily to stabilize the paste in aqueous solution before it develops adequate rigidity and, secondly, to improve the compressive strength [392,430,457]. Adding rod-like fillers to the cement formulations also caused an improvement in the mechanical properties [457]. For example, PAA and PVA were successfully used to improve the mechanical properties of a TTCP+DCPD cement but, unfortunately, with an inevitable and unacceptable reduction of both workability and setting time [458,459]. Similar findings were reported in the presence of sodium alginate and sodium polyacrylate [460]. Other polymers, such as polyphosphazene might be used as well [461-463]. Other examples of polymer/calcium orthophosphate cement formulations might be found elsewhere [464,465].

Porous calcium orthophosphate scaffolds with interconnected macropores (~ 1 mm), micropores (~ 5 μm) and of high porosity (~ 80%) were prepared by coating polyurethane foams with a TTCP+DCPA cement, followed by firing at 1200 °C. In order to improve the mechanical properties of the scaffolds, the open micropores of the struts were then infiltrated by a PLGA solution to achieve an interpenetrating bioactive ceramic/biodegradable polymer composite structure. The PLGA filled struts were further coated with a 58S bioactive glass/PLGA composite coating. The obtained complex porous biocomposites could be used as tissue engineering scaffolds for low-load bearing applications [466]. A more complicated construction, in which the PLGA macroporous phase has been reinforced with a bioresorbable TTCP+DCPA cement, followed by surface coating of the entire construct by a non-stoichiomentic CDHA layer, has been designed as well [467]. The latter approach has culminated in a unique, three-phase biocomposite that is simple to fabricate, osteoconductive and completely biodegradable.

A porosity level of 42–80% was introduced into calcium orthophosphate cement/chitosan biocomposites by addition of the water-soluble mannitol [468]. Chitosan significantly improved the mechanical strength of the entire biocomposite [469]. A similar approach was used by other researchers who studied the effect of the addition of PLGA microparticles [470-473] (which can also be loaded with drugs or growth factors [474-476]) to calcium orthophosphate cements. These biocomposites were implanted into cranial defects of rats and a content of ~ 30 wt.% of the microparticles was found to give the best results [470], while the addition of a growth factor to the biocomposites significantly increased bone contact at 2 weeks and enhanced new bone formation at 8 weeks [476]. The *in vivo* rabbit femur implant tests showed that PLGA/calcium orthophosphate cement formulations exhibited outstanding biocompatibility and bioactivity, as well as a better osteoconduction and degradability than pure calcium orthophosphate cements [471]. Further details on calcium orthophosphate cement-based biocomposites and concretes might be found in Ref. [428, chapter "Reinforced calcium orthophosphate cements"].

4.3. NANO-CALCIUM ORTHOPHOSPHATE-BASED BIOCOMPOSITES AND NANO-BIOCOMPOSITES

Nanophase materials are the materials that have grain sizes under ~ 100 nm. They have different mechanical and optical properties if compared to the large grained materials of the same chemical composition. Namely, nanophase materials have the unique surface properties, such as an increased number of atoms, grain boundaries and defects at the surface, huge surface area and altered electronic structure, if compared to the conventional micron-sized materials. For example, nano-HA (size ~ 67 nm) has a higher surface roughness of 17 nm if compared to 10 nm for the conventional submicron size HA (~ 180 nm), while the contact angles (a quantitative measure of the wetting of a solid by a liquid) are significantly lower for nano-HA (6.1) if compared to the conventional HA (11.51). Additionally, the diameter of individual pores in a nano-HA compact is five-times smaller (pore diameter ~ 6.6 Å) than that in the conventional grain-sized HA compacts (pore diameter within 19.8–31.0 Å) [477-479]. Besides, nano-HA promotes osteoblast cells adhesion, differentiation and proliferation, osteointegration and deposition of calcium containing minerals on its surface better than microcrystalline HA; thus enhancing formation of a new bone tissue within a short period [477-479]. More to the point, nano-HA was found to cause apoptosis of the leukemia P388 cells [480].

Composites of two or more materials, in which at least one of the materials is of a nanometer-scale, are defined as *nanocomposites* [32]. Natural bone mineral is a hierarchical nanocomposite of biological origin because it consists of nano-sized blade-like crystals of biological apatite grown in intimate contact with an organic matrix rich in collagen fibers and organized in a complicated hierarchical structure [21,22, 38]. Given the fact that the major organic phase of bone is collagen, *i.e.* a natural polymer (Table 1), it is obvious that a composite of a nanophase calcium orthophosphate with a biodegradable polymer should be advantageous as bone substitution material. The inorganic nanophase would be responsible for the mechanical strength (hardness) and bioactivity, while the polymer phase would provide the elasticity. In addition, the solubility of calcium orthophosphates depends on their crystallite size (smaller crystals have a higher solubility) and on their carbonate content (higher carbonate content increases the solubility) [481]. To the author's best knowledge, among calcium orthophosphates listed in Table 3, before very recently only apatites (CDHA, HA and, perhaps, FA) have been available in the nanocrystalline state. However, very recently, nano-DCPA [482-484] and nano-MCPM [485] have been synthesized and applied to prepare nano-biocomposites with strong ionic release to combat tooth caries.

A number of investigations have been conducted recently to determine the mineralization, biocompatibility and mechanical properties of the nano-biocomposites based on various (bio)polymers and nano-HA.[6] These studies covered nano-HA/PLA [269,486-493] and its copolymer with PGA [494-496], nano-HA/collagen [497-509], nano-HA/collagen/PLA [509-517], nano-HA/collagen/PVA [518], nano-HA/collagen/alginate [519,520], nano-HA/gelatin [521-526], nano-HA/poly(hexamethylene adipamide) [527], nano-HA/PPF [528], nano-HA/polyamide [529-540], nano-HA/PVA [277,278,541-543], nano-HA/PVAP [282], nano-HA/poly(ethylene-*co*-acrylic) acid [544,545], nano-HA/chitosan [546-549], nano-HA/konjac glucomannan/ chitosan [550], nano-HA/PHEMA/PCL [551], nano-HA/PCL [323,360,552,553], nano-HA/Ti [554,555], PCL semi-interpenetrating nanocomposites [556] and many other biocompatible hybrid formulations [224,258,272,348,557-575]. Several nano-biocomposites were found to be applicable as carriers for growth factors delivery [34,576,577]. Besides, the data are available on the excellent biocompartibility of such nano-

---

[6] Unfortunately, in the majority of the already published papers it often remained unclear whether "nano-HA" represented the stoichiometric nano-HA or a non-stoichiometric nano-CDHA.

biocomposites [508]. The dispersion state of nanoparticles appears to be the critical parameter in controlling the mechanical properties of nano-biocomposites, as nanoparticles always tend to aggregate owing to their high surface energy [348].

Porous (porosity ~85%) biocomposites of nano-HA with collagen and PLA have been prepared by precipitation and freeze-drying; the nano-biocomposites did not show a pH drop upon *in vitro* degradation [510-512]. They were implanted in the radius of rabbits and showed a high biocompatibility and partial resorption after 12 weeks. Nano-HA/chitosan biocomposites with improved mechanical stability were prepared from HA/chitosan nanorods [578]. Nano-HA/PLLA biocomposites of high porosity (~90%) were prepared using thermally induced phase separation [579]. Besides, nano-HA was used to prepare biocomposites with PAA and the nanostructure of the resulting nanocrystals exhibited a core-shell configuration [580,581].

Nano-HA crystals appeared to be suitable for intraosseous implantation and offered a potential to formulate enhanced biocomposites for clinical applications [582]. Thus, the biocompatibility of chitosan in osteoblast cell culture was significantly improved by addition of nano-HA [583]. Similar finding is valid for nano-HA/polyamide biocomposites [532]. Further details on nano-HA-based biocomposites might be found in an excellent review [32]. More to the point, a more general review on nanobiomaterial applications in orthopedics is also available [584], where the interested readers are referred.

4.4. BIOCOMPOSITES WITH COLLAGEN

The main constituent of the bioorganic matrix of bones is type I collagen[7] (Table 1) with molecules about 300 nm in length. This protein is conducive to crystal formation in the associated inorganic matrix. It is easily degraded and resorbed by the body and allows good attachment to cells. Collagen alone is not effective as an osteoinductive material but it becomes osteoconductive in combination with calcium orthophosphates [586]. Both collagen type I and HA were found to enhance osteoblast differentiation [587] but combined together, they were shown to accelerate osteogenesis. However, this tendency is not so straightforward: the data are available that implanted HA/collagen

---

[7] The structural and biochemical properties of collagens have been widely investigated and over 25 collagen subtypes have been identified [585].

biocomposites enhanced regeneration of calvaria bone defects in young rats but postponed the regeneration of calvaria bone in aged rats [588]. Finally, addition of calcium orthophosphates to collagen sheets was found to give a higher stability and an increased resistance to 3D swelling compared to the collagen reference [589]. Therefore, a bone analogue based on these two constituents should possess the remarkable properties. Furthermore, addition of bone marrow constituents gives osteogenic and osteoinductive properties to calcium orthophosphate/collagen biocomposites [1].

The unique characteristics of bones are the spatial orientation between the calcium orthophosphate nanophase and collagen macromolecules at the nanolevel [35], where nanocrystals (about 50 nm length) of biological apatite are aligned parallel to the collagen fibrils [21,22,31,38], which is believed to be the source of the mechanical strength of bones. The collagen molecules and the nanocrystals of biological apatite assembled into mineralized fibrils are approximately 6 nm in diameter and 300 nm long [31,35,38,511,590]. Although the complete mechanisms involved in the bone building strategy are still unclear, the strengthening effect of apatite nanocrystals in calcified tissues might be explained by the fact that the collagen matrix is a load transfer medium and thus transfers the load to the intrinsically rigid inorganic nanocrystals. Furthermore, nanocrystals of biological apatite located in between tangled fibrils cross-link the fibers either through a mechanical interlocking or by forming calcium ion bridges, thus increasing deformation resistance of the collagenous fiber network [591].

When calcium orthophosphates are combined with collagen in a laboratory, the biocomposites appear to be substantially different from natural bone tissue due to a lack of real interaction between the two components, *i.e.* interactions that are able to modify the intrinsic characteristics of the singular components themselves. The main characteristics of the route, by which the mineralized hard tissues are formed *in vivo*, is that the organic matrix is laid down first and the inorganic reinforcing phase grows within this organic matrix [21,22,31,38]. Although to date, neither the elegance of the biomineral assembly mechanisms nor the intricate composite nano-architectures have been duplicated by non-biological methods, the best way to mimic bone is to copy the way it is formed, namely by nucleation and growth of CDHA nanocrystals from a supersaturated solution both onto and within the collagen fibrils [592-594]. Such syntheses were denoted as "biologically inspired" which means they reproduce an ordered pattern and an environment very similar to natural ones [595-597]. The biologically inspired biocomposites of collagen and calcium orthophosphates (mainly, apatites) for bone substitute have a long history [29,365,500,598-616] and started from the pioneering study by Mittelmeier and Nizard [617], who mixed calcium orthophosphate granules with a collagen web. Such combinations were found to be bioactive, osteoconductive, osteoinductive [29,586,618-620] and, in general, artificial grafts manufactured from this type of the biocomposites are likely to behave similarly to bones and be of more use in surgery than those prepared from any other materials. Indeed, some data are available on the superiority of calcium orthophosphate/collagen biocomposite scaffolds over the artificial polymeric and calcium orthophosphate bioceramic scaffolds individually [621].

It has been found that calcium orthophosphates may be successfully precipitated onto a collagen substrate of whatever form or source [29,36,500,622,623]. However, adherence of calcium orthophosphate crystals to collagen did depend on how much the collagen had been denatured: the more fibrillar the collagen, the greater attachment. Clarke *et al.* first reported the production of a biocomposite produced by precipitation of DCPD onto a collagen matrix with the aid of phosphorylated amino acids commonly associated with fracture sites [603]. Apatite cements (DCPD+TTCP) have been mixed with a collagen suspension, hydrated and allowed to set. CDHA crystals were found to nucleate on the collagen fibril network, giving a material with the mechanical properties weaker than those reported for bone. More to the point, these biocomposites were without the nanostructure similar to that of bone [600,624]. The oriented growth of OCP crystals on collagen was achieved by an experimental device in which Ca^{2+} and PO_4^{3-} ions diffused into a collagen disc from the opposite directions [623,625,626]. Unfortunately, these experiments were designed to simulate the mechanism of *in vivo* precipitation of biological apatite only; due to this reason, the mechanical properties of the biocomposites were not tested [627].

Conventionally, collagen/calcium orthophosphate biocomposites can be prepared by blending or mixing of collagen and calcium orthophosphates, as well as by biomimetic methods [29,32,34,37,497,500,511,577,590,595-597,600,622,628-634]. Besides, collagen might be incorporated into calcium orthophosphate cements [600,624,635]. Typically, the type I collagen sponge is presoaked in PO_4^{3-}-containing a highly basic aqueous solution and then is immersed into a Ca^{2+}-containing solution to allow mineral deposition. Also, collagen I fibers might be dissolved in acetic acid and then this solution is added to phosphoric acid, followed by the neutralization synthesis (performed at 25 °C and solution pH within 9 – 10) between an aqueous suspension of $Ca(OH)_2$ and the H_3PO_4/collagen solution [595,596]. To ensure the quality of the final product, it is necessary to control the Ca/P ionic ratio in the reaction solution. One way to do this is to dissolve a commercial calcium orthophosphate in an acid; another one is to add Ca^{2+} and PO_4^{3-} ions in a certain ratio to the solution and after that induce the reaction [35]. Biomimetically, one can achieve an oriented growth of CDHA crystals onto dissolved collagen fibrils in aqueous solutions via a self-organization mechanism [629]. A number of authors produced calcium orthophosphate/collagen biocomposites by mixing preformed ceramic particles with a collagen suspension [636-638]. However, in all blended composites, the crystallite sizes of calcium orthophosphates were not uniform and the crystals were often aggregated and randomly distributed within a fibrous matrix of collagen. Therefore, no structural similarity to natural bone was obtained and only a compositional similarity to that of natural bone was achieved. Crystallization of CDHA in aqueous solutions might be performed in the presence of a previously dispersed collagen [29,500]. More to the point, collagen might be first dispersed in an acidic solution, followed by addition of calcium and orthophosphate ions and then co-precipitation of collagen and CDHA might be induced by either increasing the solution pH or adding mixing agents [37]. Although it resulted in biocomposites with poor mechanical properties, pressing of the HA/collagen mixtures at 40 °C under 200 MPa for several days is also known [639]. Attempts have been performed for a computer simulation of apatite/collagen composite formation process [640]. It is interesting to note, that collagen/HA biocomposites were found to possess some piezoelectric properties [641].

As the majority of the collagen/HA biocomposites are conventionally processed by anchoring micro-HA particles into collagen matrix, it makes quite difficult to obtain a uniform and homogeneous composite graft. Besides, such biocomposites have inadequate mechanical properties; over and above, the proper pore sizes have not been achieved either. Further, microcrystalline HA, which is in contrast to nanocrystalline natural bone apatite, might take a longer time to be remodeled into a new bone tissue upon the implantation. In addition, some of the biocomposites exhibited very poor mechanical properties, probably due to a lack of strong interfacial bonding between the constituents. The aforementioned data clearly demonstrate that the chemical composition similar to bone is insufficient for manufacturing the proper bone grafts; both the mechanical properties and mimetic of the bone nanostructure are necessary to function as bone in recipient sites. There is a chance for improving osteointegration by reducing the grain size of HA crystals by activating of ultrafine apatite growth into the matrix. This may lead to enhance the mechanical properties and osteointegration with improved biological and biochemical affinity to the host bone. Besides, the unidirectional porosity was found to have a positive influence on the ingrowth of the surrounding tissues into the pores of collagen/HA biocomposites [642].

Bovine collagen might be mixed with HA and such biocomposites are marketed commercially as bone-graft substitutes those further can be combined with bone marrow aspirated from the iliac crest of the site of the fracture. Application of these materials was compared with autografts for the management of acute fractures of long bones with defects, which had been stabilized by internal or external fixation [643,644]. These biocomposites are osteogenic, osteoinductive and osteoconductive; however, they lack the structural strength and require harvest of the patient's bone marrow. Although no transmission of diseases has been recorded yet, the use of bovine collagen might be a source of concern [2].

Collagen sponges with an open porosity (30–100 μm) were prepared by a freeze-drying technique and then their surface was coated by a 10-μm layer of biomimetic apatite precipitated from simulated body fluid [645]. The researchers found a good *in vitro* performance with fibroblast cell culture. Collagen/HA microspheres or gel beads have been prepared in the intention of making injectable bone fillers [646,647]. Liao *et al.* succeeded in mimicking the bone structure by blending carbonateapatite with collagen [648]. A similar material (mineralized collagen) was implanted into femur of rats and excellent clinical results were observed after 12 weeks [649]. Collagen/HA biocomposites were prepared and their mechanical performance was increased by cross-linking the collagen fibers with glutaraldehyde [501,503,504]. These biocomposites were tested in rabbits and showed a good biological performance, osteoconductivity and biodegradation. A similar approach was selected to prepare HA/collagen microspheres (diameter ~ 5 μm) by a water-oil emulsion technique in which the surface was also cross-linked by glutaraldehyde [647]. That material showed a good *in vitro* performance with osteoblast cell culture. A porous bone graft substitute was formed from a nano-HA/collagen biocomposite combined with PLA by a freeze-drying method; the resulting material was found to mimic natural bone at several hierarchical levels [511]. Subsequent *in vitro* experiments confirmed a good adhesion, proliferation and migration of osteoblasts into this composite [510]. A further increase in biocompatibility might be achieved by the addition of silicon; thus, to enhance bone substitution, Si-substituted HA/collagen composites have been developed with silicon located preferentially in the collagen phase [502]. Porous (porosity level ~ 95% with interconnected pores of 50–100 μm) biocomposites of collagen (crosslinked with glutaraldehyde) and β-TCP have been prepared by a freeze-drying technique, followed by sublimation of the solvent; the biocomposites showed a good biocompatibility upon implantation in the rabbit jaw [650].

Biocomposites of calcium orthophosphates with collagen were found to be useful for drug-delivery purposes [520,608,651-653]. Namely, an HA/collagen–alginate (20 µl) with the rh-BMP2 (100 µg/ml, 15 µl) showed bone formation throughout the implant 5 weeks after implantation without obvious deformation of the material [520]. Gotterbarm et al. developed a two-layered collagen/β-TCP implant augmented with chondral inductive growth factors for repair of osteochondral defects in the trochlear groove of minipigs. This approach might be a new promising option for the treatment of deep osteochondral defects in joint surgery [652].

To conclude this part, one should note that biocomposites of apatites with collagen are a very hot topic of the research and up to now, just a few papers are devoted to biocomposites of other calcium orthophosphates with collagen [652,654]. These biomaterials mimic natural bones to some extent, while their subsequent biological evaluation suggests that they are readily incorporated into the bone metabolism in a way similar to bone remodeling, instead of acting as permanent implant [511,617]. Collagraft®, Bio-Oss® and Healos® are several examples of the commercially available calcium orthophosphate/collagen bone grafts for clinical use [32]. However, the performance of these biocomposites depends on the source of collagen from which it was processed. Several attempts have been made to simulate the collagen-HA interfacial behavior in real bone by means of crosslinking agents such as glutaraldehyde [501,503,504,622,647,650] with the purpose to improve the mechanical properties of these biocomposites. Unfortunately, a further progress in this direction is restricted by a high cost, difficulty to control cross-infection, a poor definition of commercial sources of collagens, as well as by a lack of an appropriate technology to fabricate bone-resembling microstructures. Further details on calcium orthophosphate/collagen composites, including the list of the commercially available products, might be found elsewhere [32,612].

4.5. Biocomposites with Other Bioorganic Compounds and Biological Macromolecules

Besides collagen, both human and mammalian bodies contain dozens types of various bioorganic compounds, proteins and biological macromolecules. The substantial amounts of them potentially might be used to prepare biocomposites with calcium orthophosphates. For example, a biologically strong adhesion (to prevent invasion of bacteria) between teeth and the surrounding epithelial tissues is attributed to a cell-adhesive protein, laminin [655]. In order to mimic the nature, a laminin/apatite biocomposite layer was successfully created on the surface of both titanium [656] and EVOH [657,658] using the biomimetic approach.

Calcium orthophosphate/gelatin biocomposites are widely investigated as potential bone replacement biomaterials [255,273-275,367-375,386-393,403,430-432,457,521-526,659-670]. For example, gelatin foams were successfully mechanically reinforced by HA and then crosslinked by a carbodiimide derivative [255]. Such foams were shown to be a good carrier for antibiotic tetracycline [663]. Several biocomposites of calcium orthophosphates with alginates[8] have been prepared [390,519,520,524,596,671]. For example, porous HA/alginate composites based on hydrogels were prepared both biomimetically [596] and by using a freeze-drying technique [671]. Another research group succeeded in preparation of biphasic but monolithic scaffolds using a similar preparation route [672]. Their biocompatibility in cell culture experiments and *in vitro* biodegradability were high; however, a mechanical strength could be better.

Various biocomposites of calcium orthophosphates with chitosan [240,398,413,420,436,468,546-550,567,568,578,583,664,670,673-684] and chitin [184,395,514,685-689] are also very popular. For example, a solution-based method was developed to combine HA powders with chitin, in which the ceramic particles were uniformly dispersed [685,686]. Unfortunately, it was difficult to obtain the uniform dispersions. The mechanical properties of the final biocomposites were not very good; due to a poor adhesion between the filler and the matrix both the tensile strength and modulus were found to decrease with increase of the HA amount. Microscopic examination revealed that HA particles were intervened between the polymer chains, weakening their interactions and decreasing the entire strength [685,686].

Biocomposites of CDHA with water-soluble proteins, such as bovine serum albumin (BSA), might be prepared by a precipitation method [464,690-693]. In such biocomposites, BSA is not strongly fixed to solid CDHA, which is useful for a sustained release. However, this is not the case if a water/oil/water interfacial reaction route has been used [251]. To extend this

---

[8] Alginates are a family of unbranched binary copolymers with a structure comprising 1-4 glycosidically linked β-D-mannuronic acid and its C-5 epimer α-Lguluronic acid [596].

subject, inclusion of DNA into CDHA/BSA biocomposites was claimed [251,694-696]. Besides, bionanocomposites of an unspecified calcium orthophosphate with DNA were prepared as well [697].

Akashi and co-workers developed a procedure to prepare calcium orthophosphate-based biocomposites by soaking hydrogels in supersaturated by Ca^{2+} and PO_4^{3-} ions solutions in order to precipitate CDHA in the hydrogels (up to 70% by weight of CDHA could be added to these biocomposites) [698]. This procedure was applied to chitosan; the 3D shape of the resulting biocomposite was controlled by the shape of the starting chitosan hydrogel [699]. Another research group developed biocomposites based on *in situ* calcium orthophosphate mineralization of self-assembled supramolecular hydrogels [700].

Various biocomposites of CDHA with glutamic and aspartic amino acids, as well as poly-glutamic and poly-aspartic amino acids have been prepared and investigated by Bigi *et al.* [280,281,701-704]. These (poly)amino acids were quantitatively incorporated into CDHA crystals, provoking a reduction of the coherent length of the crystalline domains and decreasing the crystal sizes. The relative amounts of the (poly)amino acid content in the solid phase, determined through HPLC analysis, increased with their concentration in solution up to a maximum of about 7.8 wt.% for CDHA/aspartic acid and 4.3 wt.% for CDHA/glutamic acid biocomposites. The small crystal dimensions, which implied a great surface area, and the presence of (poly)amino acids were suggested to be relevant for possible application of these biocomposites for hard tissues replacement [280,281,701-704].

Recently, BCP (HA+β-TCP)/agarose macroporous scaffolds with controlled and complete interconnection, high porosity, thoroughly open pores and tailored pore size were prepared for tissue engineering application [705,706]. Agarose, a biodegradable polymer, was selected as the organic matrix because it was a biocompatible hydrogel, which acted as gelling agent leading to strong gels and fast room temperature polymerization. Porous scaffolds with the designed architecture were manufactured by combining a low temperature shaping method with stereo-lithography and two drying techniques. The biocompatibility of this BCP/agarose system was tested with mouse L929 fibroblast and human Saos-2 osteoblast during different colonization times [705].

Fibrin sealants are non-cytotoxic, fully resorbable, biological matrices that simulate the last stages of a natural coagulation cascade, forming a structured fibrin clot similar to a physiological clot [707]. Biocomposites of calcium orthophosphates with fibrin sealants might develop the clinical applications of bone substitutes. The 3D mesh of fibrin sealant interpenetrates the macro- and micro-porous structure of calcium orthophosphate ceramics [9]. The physical, chemical and biological properties of calcium orthophosphate bioceramics and the fibrin glue might be cumulated in biocomposites, suitable for preparation of advanced bone grafts [708-719].

Furthermore, there are biocomposites of calcium orthophosphates with bisphosphonates [720], silk fibroin (that is a hard protein extracted from silk cocoon) [250,563-565,570,571,721-726], chitosan + silk fibroin [727], fibronectin [728] and casein phosphopeptides [729]. Besides, the reader's attention is pointed out to an interesting approach to crystallize CDHA inside poly(allylamine)/poly(styrene sulfonate) polyelectrolyte capsules resulting in empty biocomposite spheres of micron size [730]. Depending on the amount of precipitated CDHA, the thickness of the shell of biocomposite spheres can be varied between 25 and 150 nm. These biocomposite capsules might find application as medical agents for bone repairing and catalytic microreactors [730].

4.6. Injectable Bone Substitutes (IBS)

Injectable bone substitutes (IBS) represent ready-to-use suspensions of calcium orthophosphate powder(s) in a liquid carrier phase. They look like viscous pastes with the rheological properties, sufficient to inject them into bone defects by means of surgical syringes and needles. Usually, the necessary level of viscosity is created by addition of water-soluble polymers [104,731,732]. Therefore, the majority of calcium orthophosphate-based IBS formulations might be considered as a subgroup of calcium orthophosphate/polymer biocomposites. For example, an IBS was described that involved a silanized hydroxyethylcellulose carrier with BCP, consisting of HA and β-TCP [733]. The suspension is liquid at pH within 10-12, but gels quickly at pH < 9. Injectable composites can be formed with β-TCP to improve mechanical integrity [454]. Similarly, Bennett *et al.* showed that a polydioxanone-*co*-glycolide-based biocomposite reinforced with HA or β-TCP can be used as an injectable or moldable putty [734]. During the crosslinking reaction following injection, carbon dioxide is released allowing the formation of interconnected pores.

Daculsi *et al.* developed viscous IBS biocomposites based on BCP (60% HA + 40% β-TCP) and 2% aqueous solution of HPMC that was said to be perfectly biocompatible, resorbable and easily fitted bone defects (due to an initial plasticity) [84,732,735-741]. The best ratio BCP/HPMC aqueous solution was found to be at ~ 65/35 w/w. To extend this subject further, this type of IBS might be loaded by cells [742] or by microparticles [743].

The advanced characteristics of IBS come from their good mechanical properties and biocompatibility and the ease of tissue regeneration. Although the fabrication of IBS biocomposites in most cases improved the mechanical properties of the system and provided the material with resistance to fluids penetration, these achievements were limited by the amount of polymer that can be added to the paste. For instance, Mickiewicz *et al.* reported that after a critical concentration (that depended on the type and molecular weight of the polymer, but was always around 10%), the polymer started forming a thick coating on the crystal clusters, preventing them from interlocking, originating plastic flow and, as a consequence, decreasing mechanical properties [464]. More to the point, Fujishiro *et al.* reported a decrease in mechanical properties with higher amounts of gel, which was attributed to formation of pores due to leaching of gelatin in solution [457]. Therefore, it seems that mechanical properties, although improved by the addition of polymers, are still a limitation for the application of calcium orthophosphate-based IBS formulations in load-bearing sites [147].

4.7. BIOCOMPOSITES WITH GLASSES, INORGANIC MATERIALS AND METALS

To overcome the problem of poor mechanical properties of calcium orthophosphate bioceramics, suitable biocomposites of calcium orthophosphates reinforced by various inorganic materials, glasses and metals have been developed. Such biocomposites are mainly prepared by the common ceramic processing techniques such as thermal treatment after kneading [744-746], powder slurry coating [747] and metal-sol mixing [748]. For example, HA was combined with Bioglass® (Novabone Products, Alachua, FL) [749,750] and with other glasses [751] to form glass-ceramics biocomposites. Other reinforcement materials for calcium orthophosphates are differentiated by either shape of the fillers, namely, particles [752,753], platelets [754,755], whiskers [485,756,757], fibers [758-760], or their chemical composition: zirconia and/or PSZ [251,744-747,756,761-794], alumina [251,752,755,794-803], titania [308,748,753,804-818], other oxides [819-822], silica and/or glasses [823-830], wollastonite [172,831-838], various metals and alloys

[760,795,818,839-852], calcium sulfate [853-855], silicon carbide [757], barium titanate [856], zeolite [857] and several other materials [272,858-860]. All these materials have been added to calcium orthophosphate bioceramics to improve its reliability. Unfortunately, significant amounts of the reinforcing phases are needed to achieve the desired properties and, as these materials are either bioinert, significantly less bioactive than calcium orthophosphates or not bioresorbable, the ability of the biocomposites to form a stable interface with bone is poorer if compared with calcium orthophosphate bioceramics alone. Due to the presence of bioinert compounds, such formulations might be called bioinert/bioactive composites [823]. The ideal reinforcement material would impart mechanical integrity to a biocomposite at low loadings, without diminishing its bioactivity. As clearly seen from the amount of the references, apatite/zirconia biocomposites are most popular ones among the researchers.

There are several types of HA/glass biocomposites. The first one is also called bioactive glass-ceramics. A dense and homogeneous biocomposite was obtained after a heat treatment of the parent glass, which comprised ~38 wt.% oxy-FAP ($Ca_{10}(PO_4)_6(O,F)_2$) and ~34 wt.% β-wollastonite ($CaO \cdot SiO_2$) crystals, 50–100 nm in size in a $MgO-CaO-SiO_2$ glassy matrix [172,831-838]. Apatite-wollastonite (A-W) glass-ceramics is an assembly of small apatite particles effectively reinforced by wollastonite. The bending strength, fracture toughness and Young's modulus of A-W glass-ceramics are the highest among bioactive glass and glass ceramics, enabling it to be used in some major compression load bearing applications, such as vertebral prostheses and iliac crest replacement. It combines a high bioactivity with the suitable mechanical properties [861]. β-TCP/wollastonite biocomposites are also known [862-864]. More complicated biocomposites have been developed as well. For example, (A-W)/HDPE composite (AWPEX) biomaterials have been designed to match the mechanical strength of human cortical bone and to provide favorable bioactivity, with potential use in many orthopedic applications [865-868]. Other examples comprise wollastonite-reinforced HA/Ca polycarboxylate [869] and glass-reinforced HAP/polyacrylate [870] biocomposites.

HA/glass biocomposites can be prepared by simple sintering of appropriate HA/glass powder mixtures [871-874]. If sintering is carried out below 1000 °C, HA does not react with the bioactive glass [872,873] or this reaction is limited [874]. Besides, reaction between HA and glasses depends on the glass composition. In another approach, small quantities of bioactive glass have been added to HA bioceramics in order to improve densification and/or mechanical properties [26]. In addition, biocomposites might be sintered from HA and silica [823]. In general, bioactive glass-ceramics maintain a high strength for a longer time than HA bioceramics under both the *in vitro* and *in vivo* conditions [830,835].

Carbon nanotubes with their small dimensions, a high aspect ratio (length to diameter) as well as the exceptional mechanical properties, including extreme flexibility and strength, significant resistance to bending, high resilience and the ability to reverse any buckling of the tube, have the excellent potential to accomplish necessary mechanical properties [875]. Recent studies have even suggested that they may possess some bioactivity [876-879]. However, due to a huge difference in shapes, it is a challenge to prepare homogeneous mixtures of calcium orthophosphates and carbon nanotubes: "one can imagine something similar to achieving a homogeneous mixture of peas and spaghetti" [875, page 7]. Additionally, non-functionalized carbon nanotubes tend to agglomerate and form bundles; besides, they are soluble in neither water nor organic solvents. Chemical functionalization allows carbon nanotubes to be dispersed more easily, which can improve interfacial bonding with calcium orthophosphates [248,875].

Different strategies might be employed to prepare calcium orthophosphate/carbon nanotubes biocomposites. For example, apatites might be chemically synthesized by using carboxyl functionalized carbon nanotubes as a matrix [243-248]. Physico-chemical characterization of these biocomposites showed that nucleation of CDHA initiates through the carboxyl group [248]. Hot-pressing [880], plasma spraying [881] and laser surface alloying [882-884] techniques might be applied as well. The research on calcium orthophosphate (up to now, only apatites)/carbon nanotube biocomposites is in its early stages, with the first papers published in 2004 [247,434]. Due to this reason, the mechanical property data for such biocomposites have been reported only in few papers; however, these results are encouraging. For example, Chen *et al.* performed nanoindentation tests on biocomposite coatings to give hardness and Young's modulus values [884]. They found that the higher the loading of nanotubes, the better the properties. Namely, at 20 wt.% loading, hardness was increased by 43% and Young's modulus by 21% over a single-phase HA coating [884]. Scratching test results indicated that as alloyed HA biocomposite coatings exhibited improved wear resistance and lower friction coefficient with increasing the amount of carbon nanotubes in the precursor material powders [883]. Additionally, measurements of the elastic modulus and hardness of the biocomposite coatings indicated that the mechanical properties were also affected by the amount of carbon nanotubes [882]. Another research group performed compression tests on bulk HA/nanotubes biocomposites and found an increase in strength over single-phase HA [247]. However, the highest compressive strength they achieved for any material was only 102 MPa, which is similar to that of cortical bone but much lower than the typical values for dense HA [875]. More complex formulations, such as poly-L-lysine/HA/carbon nanotube hybrid nanocomposites, have been also developed [885]. Unfortunately, carbon nanotubes are very stable substances; they are neither bioresorbable nor biodegradable. Therefore, during the *in vivo* bioresorption, the nanotubes will get into the human body from the biocomposite matrix and might cause uncertain health problems. Except of carbon nanotubes, carbon fibers of microscopic dimensions are also used to reinforce HA bioceramics [886-888].

The main disadvantage of HA reinforced by PSZ is degradation of zirconia in wet environments [756,761,762,784]. Transformation of the tetragonal ZrO_2 to the monoclinic phase on the surface results in formation of microcracks and consequently lowers the strength of the implant [889,890].

An HA-based biocomposite reinforced with 20 vol. % of Ti particles was fabricated by hot pressing [841]. Besides, calcium orthophosphates/Ti biocomposites might be prepared by powder metallurgy processing [843-845]. At high temperatures, the presence of Ti metal phase was found to promote dehydration and decomposition of HA into β-TCP and TTCP [841,843] or partial formation of β-TCP and calcium titanate instead of HA [555,844,845]. Comparing with pure HA bioceramics manufactured under the same conditions, the HA/Ti biocomposites possessed a higher fracture toughness, bending strength, work of fracture, porosity and lower elastic modulus, which is more suitable for biomedical applications. However, the mechanical properties appeared to be not high enough to use HA/Ti biocomposites in load-bearing applications. Luckily, the histological evaluations revealed that HA/Ti biocomposites could be partially integrated with newborn bone tissues after 3 weeks and fully osteointegrated at 12 weeks *in vivo* [841]. Similar findings had been earlier made for HA bioceramics reinforced by addition of silver particulates (5–30 vol. %) and subsequent sintering of the HA/Ag powder compacts [839,840]. Other studies on calcium orthophosphate/Ti biocomposites are available elsewhere [846-849].

To conclude this part, biocomposites consisting of calcium orthophosphates only should be briefly described. First of all, BCP itself, consisting of HA and α- or β-TCP, should be mentioned [23]. In 1980-s, BCP was called as "TCP ceramics complexed with HA" [891]. More to the point, 70% HA-powder + 30% HA-whisker biocomposites have been fabricated by pressureless sintering, hot pressing and hot isostatic pressing. These biocomposites were found to exhibit an improved toughness, attaining the lower fracture-toughness limit of bone without a decrease of bioactivity and biocompatibility [892,893]. Besides, a dual HA biocomposite that combined two HA materials with different porosities: HA with 75% porosity, for bone ingrowth and HA with 0% porosity, for load bearing was manufactured. This dual HA biocomposite appeared to be suitable for use as an implant material for spinal interbody fusion as a substitute for iliac bone grafts, which could eliminate the disadvantages associated with autograft harvesting [894]. A biodegradable nanocomposite porous scaffold comprising a β-TCP matrix and HA nanofibres was developed and studied for load-bearing bone tissue engineering. HA nanofibres were prepared by a biomimetic precipitation method, the inclusion of which significantly enhanced the mechanical property of the scaffold, attaining a compressive strength of 9.87 MPa, comparable to the high-end value (2–10 MPa) of cancellous bone [895].

4.8. FUNCTIONALLY GRADED BIOCOMPOSITES

Although, in most cases, the homogeneous distribution of filler(s) inside a matrix is required [356], there are composites, where this is not the case. For example, functionally graded materials (commonly referred to as FGM) might be characterized by the intentional variations in composition and/or structure gradually over volume, resulting in corresponding changes in the properties of the composite. The main feature of such materials is the almost continuously graded composition that results in two different properties at the two ends of the structure. Such composites can be designed for specific function and applications. Various approaches based on the bulk (particulate processing), preform processing, layer processing and melt processing are used to fabricate the functionally graded materials.

Bone is a biologically formed composite with variable density ranging from very dense and stiff (the cortical bone) to a soft and foamed structure (the trabecular bone). Normally the outer part of long bones consists of cortical bone with the density decreasing towards the core, where the trabecular bone is found. The trabecular bone is porous and the porosity is filled with osseous medulla [21,22]. This brief description clearly indicates that bones are natural functionally graded composites.

The concept of FGM has been increasingly used for biomaterial design and currently it remains to be an important area of the research. For example, powder metallurgy methods have been used to fabricate HA/Ti functionally graded biocomposite dental implants offering the biocompatible HA on the tissue side and titanium on the outer side for mechanical strength [896-898]. The graded structure in the longitudinal direction contains more Ti in the upper section and more HA in the lower section. Actually, in the upper section the occlusal force is directly applied and Ti offers the required mechanical performance; in the lower part, which is implanted inside the bone, the HA confers the bioactive and osteoconductive properties to the material [896]. Since the optimum conditions of sintering for Ti and HA are very different, HA/Ti functionally graded biocomposites are difficult to fabricate and the sintering conditions for their mixtures are obliged to compromise. The expected properties of this implant are shown in Fig. 3 [897]. Functionally graded HA/Ti biocomposite coatings might be prepared by rf-plasma spraying [899]. A functionally graded HA/PMMA biocomposite was developed based on sedimentary HA distributions in a PMMA viscous fluid, using a centrifuge to avoid stress convergence on the interface. The stress-strain curves of this biocomposite showed sufficient strength for medical application along with the relaxation of brittleness and fragility [449]. A three-layered graded biocomposite membrane, with one face of 8% nano-carbonated CDHA/collagen/PLGA porous membrane, the opposite face of pure PLGA non-porous membrane, the middle layer of 4% nano-carbonated CDHA/collagen/PLGA as the transition was prepared through the layer-by-layer casting method [513]. HA/glass FGM layers were coated on titanium alloy (Ti–6Al–4V) substrates. The design of these layers and the use of the glass were for achieving a strong bonding between the FGM layered coatings and the substrates [900,901]. More to the point, Ti alloy substrate has been combined with HA granules spread over the surface [902].

Functionally graded β-TCP/FA biocomposites combine the biostability of FA with bioresorbable properties of β-TCP [903]. An interesting multilayered (each layer of 1 mm thick) structure consisting of β-TCP/FA biocomposites with different molar ratios has been prepared, giving rise to formation of a FGM (Fig. 4). After implantation, the preferential dissolution of β-TCP phase would result in functionally gradient porosity for bone ingrowth [904]. HA/zirconia graded biocomposites were fabricated to enhance the mechanical properties of HA while retaining its bone bonding property [792]. TiO_2 and HA were found to be a good combination for FGM providing both a gradient of bioactivity and a good mechanical strength [904]. Besides, graded HA/$CaCO_3$ biocomposite structures for bone ingrowth have been developed as well [905]. Functionally graded composite skull implants consisting of polylactides, carbonateapatite and $CaCO_3$ are known as well [319,320]. The research in this field is quite promising but currently the mechanical properties of the available biocomposites are clearly in excess of the properties of bone [148].

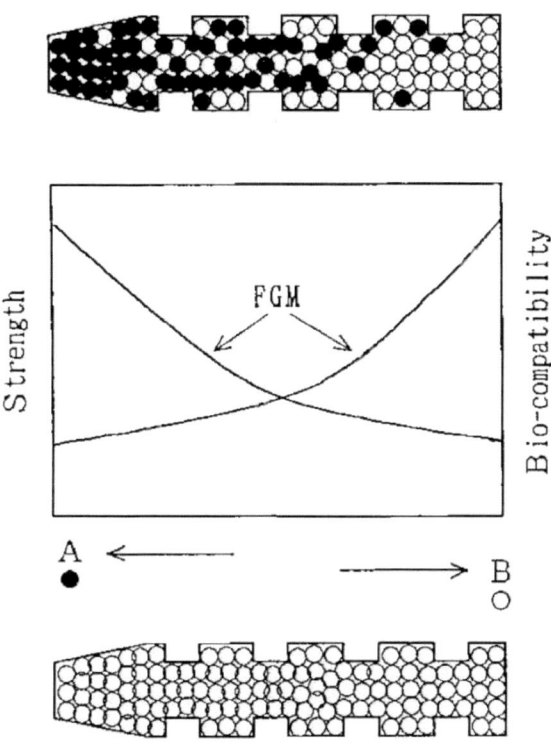

Figure 3. Expected properties of functionally graded biocomposite dental implant. For comparison, the upper drawing shows a functionally graded implant and the lower one shows a conventional uniform implant. The properties are shown in the middle. The implant with the composition changed from a biocompatible metal (Ti) at one end (left in the figure), increasing the concentration of bioceramics (HA) toward 100% HA at the other end (right in the figure), could control both mechanical properties and biocompatibility without an abrupt change due to the formation of discrete boundary. This FGM biocomposite was designed to provide more titanium for the upper part where occlusal force is directly applied and more HA for the lower part, which is implanted inside the jawbone. Reprinted from Ref. [897] with permission.

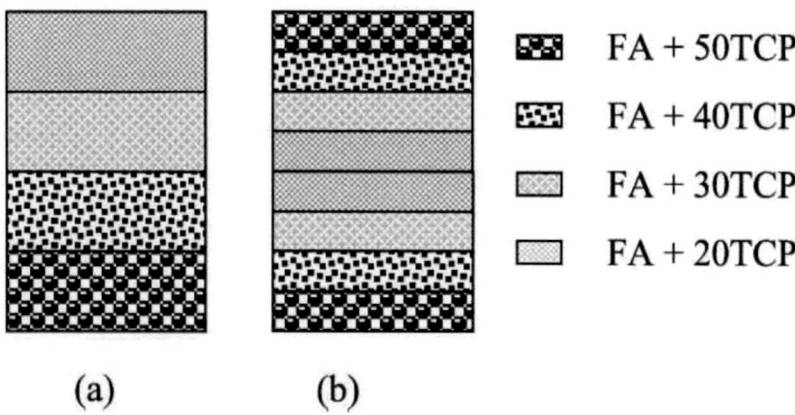

Figure 4. A schematic diagram showing the arrangement of the FA/β-TCP composite layers: (a) non-symmetric FGM, (b) symmetric FGM. Reprinted from Ref. [903] with permission.

4.9. BIOSENSORS

A biosensor is a device for detection of an analyte that combines a biological component with a physicochemical detector component. Very briefly, it consists of 3 parts: a sensitive biological element; a transducer or a detector element that transforms the signal resulting from the interaction of the analyte with the biological element into another signal; and associated electronics that is primarily responsible for the display of the results in a user-friendly way [906].

The surface of biologically relevant calcium orthophosphates (CDHA, HA, α-TCP, β-TCP) has an excellent ability of adsorption for functional biomolecules such as proteins, albumins, DNA and so on. Therefore, some calcium orthophosphate-based biocomposites and hybrid biomaterials were found to be applicable for biosensor manufacturing [289,543,852,885]. For example, formation of poly-L-lysine/HA/carbon nanotube hybrid nanoparticles was described and a general design strategy for an immunosensing platform was proposed based on adsorption of antibodies onto this nanocomposite [885]. In another paper, a hybrid material formed by assembling of gold nanoparticles onto nano-HA was employed for the interface design of piezoelectric immunosensor, on which the antibodies were bound. The developed sensing interface appeared to possess some advantages, such as activation-free immobilization and high antigen-binding activities of antibodies, over using either nano-HA or gold nanoparticles alone [852]. Up to date, just a few papers have been published on biosensor application of calcium orthophosphate-based biocomposites. Presumably, this subject will be further developed in the future and, perhaps, sometime implantable biosensors will be designed to perform the continuous concentration monitoring of the important biological macromolecules. Possibly, those biocencors might be able to use an electric power, generated by DCPD/polymer composite-based battery devices [414,415].

Chapter 5

INTERACTION BETWEEN THE PHASES IN CALCIUM ORTHOPHOSPHATE-BASED BIOCOMPOSITES

An important aspect that should be addressed in details is a mutual interaction between calcium orthophosphates and other phases in biocomposites and hybrid biomaterials. In general, an interaction between the phases in any composite can be either mechanical, when it results from radial compression forces exerted by the matrix on the filler particles (for example, developed during cooling due to thermal contraction), or chemical, when the reactivity of the filler towards the matrix has an important role. In the latter case, it is important to distinguish a physical interaction from chemical bonding [226]. According to Wypych [907], physical interaction is more or less temporary, implicating hydrogen bonding or van der Waals forces, whereas chemical bonding is stronger and more permanent, involving covalent bond formation. Thus, a chemical interfacial bond between the phases is preferred to achieve a higher strength of a composite. The magnitude of the interfacial bond between the phases determines how well a weak matrix transmits stress to the strong fibers. However, while a bond between the matrix and reinforcement must exist for the purpose of stress transfer, it should not be so strong that it prevents toughening mechanisms, such as debonding and fiber pullout [875].

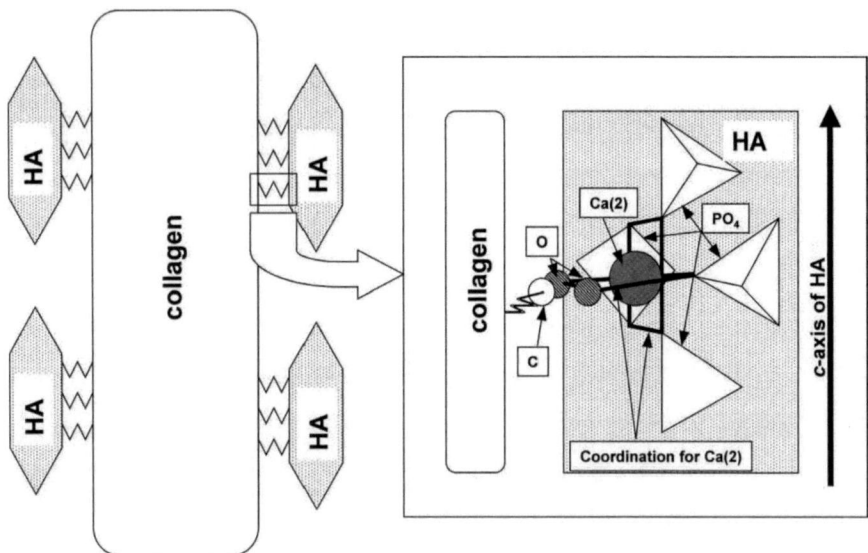

Figure 5. A schematic drawing of the relation between self-organization (directional deposition of HA on collagen) and interfacial interaction in biocomposites. Direction of interaction between HA and collagen is restricted by covalent bond between COO and Ca(2) to maintain regular coordination number of 7. Reprinted from Ref. [629] with permission.

There is still doubt as to the exact bonding mechanism between bone minerals (biological apatite) and collagen, which undoubtedly plays a critical role in determining the mechanical properties of bones. Namely, bone minerals are not directly bonded to collagen but through non-collagenous proteins that make up ~3 % of bones (Table 1) and provide with active sites for biomineralization and for cellular attachment [32]. In bones, the interfacial bonding forces are mainly ionic bonds, hydrogen bonds and hydrophobic interactions, which give the bones the unique composite behavior [49]. There is an opinion that, opposite to bones, there is no sign of chemical bonding between phases in conventional calcium orthophosphate/collagen biocomposites, probably due to a lack of suitable interfacial bonding during mixing [35]. However, this is not the case for phosphorylated collagens [634]. Anyway, Fourier-transformed infrared (FTIR) spectra of some calcium orthophosphate-based composites and collagen films were measured and transformed into absorption spectra using the Kramers-Kronig equation to demonstrate energy shifts of residues on the HA/collagen interface. After comparing FTIR spectra of biocomposites and collagen films in detail, red shifts of the absorption bands for C–O bonds were observed in the spectra of the biocomposites. These red shifts were described as a decrease of bonding energies of C–O bonds and assumed to be caused by an interaction to Ca^{2+} ions located on the surfaces of apatite nanocrystals, as shown in Fig. 5 [629]. Another proof of a chemical interaction between CDHA and collagen fibers was also evaluated in FTIR spectra of CDHA/collagen biocomposites, in which a shift of the band corresponding to –COO⁻ stretching from 1340 to 1337 cm^{-1} was observed [595,596]. More to the point, nucleation of CDHA crystals onto collagen through a chemical interaction with carboxylate groups of collagen macromolecules has been reported [908-910].

FTIR-spectroscopy seems to be the major investigation tool of a possible chemical bonding among the phases in calcium orthophosphate-based biocomposites and hybrid biomaterials [221,282,288,290,383,421,503,518,527,530,537,540,542,545,550,557,566,571 ,596, 634,667,668,727,911,912]. For example, the characteristic bands at 2918, 2850 and 1472 cm^{-1} for the hydrocarbon backbone of PE appeared to have zero shift in an HA/PE biocomposite. However, in the case of polyamide, some of the FTIR-bands indicated that the polar groups shifted apparently: the bands at 3304, 1273 and 692 cm^{-1} derived from stretching of N–H, stretching of C–N–H and vibrating of N–H moved to 3306, 1275 and 690 cm^{-1} in an HA/polyamide biocomposite, respectively. Both stretching (3568 cm^{-1}) and vibrating (692 cm^{-1}) modes of hydroxyl in HA moved to 3570 and 690 cm^{-1} in the HA/polyamide buicomposite respectively, indicating the formation of hydrogen bonds. Besides, the bands at 1094 and 1031 cm^{-1} of PO_4 modes also shifted to 1093 and 1033 cm^{-1} in the HA/polyamide biocomposite. The bands shift in a fingerprint area indicated that the hydroxyl and orthophosphate on the surface of HA might interact with plentiful carboxyl and amino groups of polyamide through nucleophilic addition [221]. Comparable conclusions were made for nano-HA/PVA [542], CDHA/alginate [596], ACP/PPF [421], HA/maleic anhydride [290] and β-TCP/PLLA [383] biocomposites, where a weak chemical bond was considered to form between Ca^{2+} ions located on the nano-HA, CDHA, ACP, HA or β-TCP surface, respectively, and slightly polarized O atoms of C=O bonds in the surrounding bioorganic compounds. Schematically, this chemical interaction is shown in Fig. 6 [596].

Except of FTIR-spectroscopy, other measurement techniques are also able to show some evidences of a chemical interaction between calcium orthophosphates and other compounds in biocomposites [282,383,537,540, 542,912-914]. For example, for CDHA/alendronate nanocrystals such evidences were observed by thermogravimetric analysis: DTG plots of the nanocrystals appeared to be quite different from those obtained from mechanical mixtures of CDHA and calcium alendronate with similar compositions [913]. Analogous DTG results were obtained for nano-HA/PVA [542]. In the case of nano-HA/polyamide biocomposites, a hydrogen bonding between the phases was detected by differential scanning calorimetry technique [537]. Another example comprises application of the dynamic mechanical analysis to investigate softening mechanism of β-TCP/PLLA biocomposites [383]. In the case of nano-HA/PVAP composites, the indirect evidences of chemical bonding between the phases were found by X-ray diffraction and thermogravimetric analysis [282]. A strong structural correlation between the orientation of FA crystallites and gelatin within the FA/gelatin composite spheres was discovered that indicated to a substantial reorganization of the macromolecular matrix within the area of a growing aggregate [367].

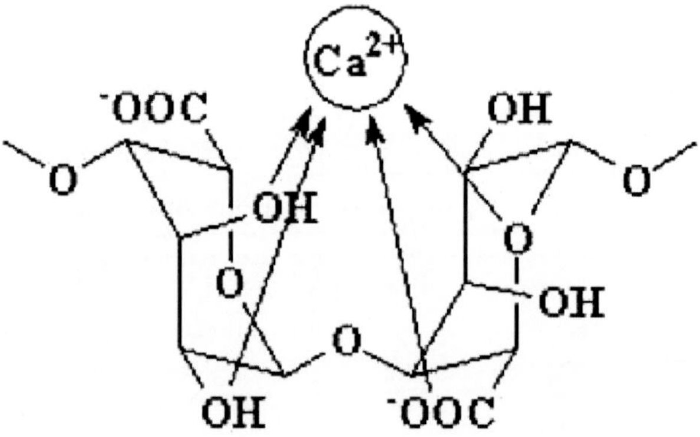

Figure 6. A schematic diagram of Ca^{2+} ion binding with alginate chains. Reprinted from Ref. [596] with permission.

By means of the X-ray photo-electronic spectroscopy (XPS) technique, binding energies of Ca, P and O atoms were found to have some differences between nano-HA (Ca: 350.5 and 345.5; O: 530.2; P: 132.5 eV) and nano-HA/konjac glucomannan/chitosan biocomposite (Ca: 352.1 and 347.4; O: 531.2; P: 133.4 eV), respectively [550]. Further measurements by FTIR and X-ray diffraction revealed that nano-HA was mainly linked with konjac glucomannan and chitosan by hydrogen bonding among OH^- and PO_4^{3-} of nano-HA and –C=O and –NH of konjac glucomannan and chitosan copolymer and there was a stable interface formed between the three phases in the biocomposite. Meanwhile, coordinate bonding might be formed between Ca^{2+} and –NH. Stable interfaces have been formed among the three phases in a biocomposite [550]. In HA/collagen biocomposites, a covalent bond formation between Ca^{2+} of HA and $RCOO^-$ of collagen molecules was found by XPS [504]. Similar XPS observations were also made for several other calcium orthophosphate-based biocomposites [530,557,566].

The interaction and adhesion between calcium orthophosphate fillers and respective matrices have a significant effect on the properties of particulate filled reinforced materials, being essential to transfer the load between the phases and thus improve the mechanical performance of the composites [288]. However, for the substantial amount of the discussed in this book biocomposites, the interaction between the phases is mechanical in nature. This is because the matrix often consists of compounds with no functional groups or unsaturated bonds, which can form ionic complexes with the constituents of calcium orthophosphates. Obviously, less coupling exists between non-polar polymers and calcium orthophosphate ceramic particles. Therefore, polymers with functional groups pendant to the polymer backbone, which can act as sites for bridging to calcium orthophosphates, are more promising in this respect [49]. Besides, the surface of calcium orthophosphates might be modified as well [116,417,418,553,915,916]. In order to improve the situation, various supplementary reagents are applied. Namely, if the primary effect of a processing additive is to increase the interaction between the phases, such an additive can be regarded as a coupling agent [917]. Coupling agents establish chemical bridges between the matrix and the fillers, promoting the adhesion between the phases. In many cases, their effect is not unique, influencing also the rheology of composites [226].

Optimization of biocomposite properties with coupling agents is currently an important area of the research. The control and development of molecular-level associations of polymer with calcium orthophosphates is suggested to be significant for the resulting mechanical responses in the composites. It appears that a fundamental molecular understanding of interfacial behavior in biocomposite systems is an area not sufficiently addressed in literature. Various experimental characterization techniques using electron microscopy, vibrational spectroscopy, X-ray diffraction, scanning probe microscopy and others are used routinely to characterize these materials besides mechanical property characterization. In addition, atomic scale models for simulating the phase interaction and predicting responses in the novel material systems, where nanostructure and nanointerfaces are included, are important to understand and predict the load deformation behavior [148].

A hexamethylene diisocyanate coupling agent was used to bind PEG/PBT (Polyactive™) block copolymers [235] and other polymers [911] to HA filler particles. Thermogravimetric and infrared analysis demonstrated that the polymers were chemically bonded to the HA particles through the isocyanate groups, making it a suitable approach to improve the adhesion [911]. Other researchers used glutaraldehyde as a cross-linked reagent in various calcium orthophosphate-based biocomposites 389,393,501,503,504,521,526,586,622, 647,650,918]. The interfacial bonding between calcium orthophosphates and other components might be induced by using various coupling agents and surface modifiers, such as silanes [193,235,338,541,919-924], zirconates [226,338,340,915,925], titanates [226,338,925], phosphoric acid [544], alkaline pretreatment [723,726], polyacids [115,116,235] and other chemicals. Besides, some polymers might be grafted onto the surface of calcium orthophosphates [553]. Structural modifications of the polymeric matrices, for instance, with introduction of acrylic acid [196,235,920,921], have also proved to be effective methods. For example, application of polyacids as a bonding agent for HA/Polyactive™ composites caused the surface modified HA particles to maintain better contact with the polymer at fracture and improved mechanical properties [115,116,235]. The use of titanate and zirconate coupling agents appeared to be very dependent on the molding technique employed [226]. Silane-coupled HA powders were tested before applying them as fillers in biodegradable composites [922-924]. This treatment allowed HA withstanding the attack of water without impairing overall bioactivity. Besides, chemically modified reinforcement phase-matrix interface were found to improve the mechanical properties of the biocomposites. Examples of such interface modified biocomposites include chemically coupled HA/PE [920,921], chemically formed HA/Ca poly(vinylphosphonate) [284] and PLA/HA fibers [185]. These biocomposites are able to consume a large amount of energy in the fracture.

The action of some coupling agents was found to combine two distinct mechanisms: (i) crosslinking of the polymeric matrix (valid for zirconate and titanate coupling agents) and (ii) improvement of the interfacial interactions between the major phases of the composites. This interfacial adhesion improvement appeared to be much dependent on the chemical nature (pH and type of metallic centre) of the coupling agents [338]. Several works claimed that silanes do interact with HA [193,920-924]. It was shown that a silicon-containing inter-phase existed between HA and PE, which promoted the chemical adhesion between the HA particles and the polymer. A silane-coupling agent also facilitated penetration of PE into cavities of individual HA particles, which resulted in enhanced mechanical interlocking at the matrix-reinforcement interface [920,921].

Addition of adhesion promoting agents might be an alternative to improve the interaction between the fillers and the matrix. For example, Morita *et al.* used incorporation of 4-methacryloyloxyethyl trimellitate anhydride to promote adhesion of the polymer to HA [926]. In another study, phosphoric ester was added to the liquid component of the formulation [927]. Both the strength and the affinity index of biocomposites were found to increase, probably due to the effects of co-polymerization.

Possible interactions between BCP and HPMC have been investigated in IBS composites [737,738,928]. After mixing, there was a decrease in the mean diameter of BCP granules and this influenced the viscosity of the paste. Dissolution of grain boundaries of β-TCP crystals and precipitation of CDHA on HA crystal surface was found during the interaction between BCP and HPMC in aqueous solutions. Both phenomena were responsible for the observed granulometric changes [737,738]; however, within the sensitivity of the employed measurement techniques, no chemical bonding between BCP and HPMC was detected [928].

A co-precipitation method was used to prepare CDHA/chitosan biocomposites [673]. Growth of CDHA crystals was inhibited by organic acids with more than two carboxyl groups, which strongly bind to CDHA surfaces via a COO–Ca bond. Transmission electron microscopy images revealed that CDHA formed elliptic aggregates with chemical interactions (probably coordination bond) between Ca on its surface and amino groups of chitosan; the CDHA nanocrystals were found to align along the chitosan molecules, with the amino groups working as the nucleation sites [673]. Formation of calcium cross-linked polymer carboxylate salts was suggested during setting of calcium orthophosphate cement (TTCP+DCPA)/polyphosphazane biocomposites; the chemical involvement of the polymer in the cement setting was concluded based on the results of pH monitoring [461-463].

A chemical bond between the phases was presumed in PCL/HA composites, prepared by the grafting technique [351]; unfortunately, no strong experimental evidences were provided. In another study, CDHA/poly(α-hydroxyester) composites were prepared by a low temperature chemical route [325]. In that study, pre-composite structures were prepared by combining α-TCP with PLA, PLGA and copolymers thereof. The final biocomposite structure was achieved by *in situ* hydrolysis of α-TCP to CDHA performed at 56 °C either in solvent cast or pressed pre-composites. That transformation occurred without any chemical reaction between the polymer and calcium orthophosphates, as it was determined by FTIR spectroscopy [325].

In nearly every study on HA/carbon nanotubes biocomposites, the nanotubes have been functionalized before combining them with HA. Most researchers have done this by oxidation [243-247], although non-covalent functionalizing with sodium dodecylsulfate [247] and coating the nanotubes by a polymer [929] before combining them with HA have also been reported. Several studies by transmission electron microscopy have shown evidences that the functionalization has enhanced interaction between carbon nanotubes and HA [246,247,930].

If calcium orthophosphate-based biocomposites are able to sustain a high-temperature sintering (valid for the formulations consisting of inorganic components only), an inter-diffusion of chemical elements will take place between the phases. Such effect has been detected by energy-dispersive X-ray spectroscopy in HA/TiO_2 biocomposite particles with partial formation of calcium titanates; this process was found to be favorable to enhancing the cohesive strength of particles in the composite coating [818]. A similar high-temperature interaction between HA and zirconia [744,769], as well as between HA and Ti [555,841,843-845] was also detected. Besides, partial decomposition of HA and formation of different calcium aluminates were detected in HA/Al_2O_3 biocomposites after sintering at 1200-1300 °C [796,802,803].

Chapter 6

BIOACTIVITY AND BIODEGRADATION OF CALCIUM ORTHOPHOSPHATE-BASED BIOCOMPOSITES

The continuous degradation of an implant causes a gradual load transfer to the healing tissue, preventing stress-shielding atrophy and stimulates the healing and remodeling of bones. Some requirements must be fulfilled by the ideal prosthetic biodegradable materials, such as biocompatibility, adequate initial strength and stiffness, retention of mechanical properties throughout sufficient time to assure its biofunctionality and non-toxicity of the degradation by-products [147]. Generally speaking, bioactivity (*i.e.*, ability of bonding to bones) of biologically relevant calcium orthophosphates reinforced by other materials is usually lower than that of pure calcium orthophosphates [27,28,931].

In general, both bioactivity and biodegradability of any biocomposite are determined by the same properties of the constituents. Both processes are very multi-factorial because, after implantation, the surface of any graft is rapidly colonized by cells. Much more biology, than chemistry and material science altogether, is involved into these very complex processes and many specific details still remain unknown. To simplify the task, the biodegradability of the biologically relevant calcium orthophosphates might be described by a chemical dissolution in slightly acidic media (calcium orthophosphates are almost insoluble in alkaline solutions [87-93]), which, in the case of CDHA, might be described as a sequence of four successive chemical equations [428,932,933]:

$$Ca_{10-x}(HPO_4)_x(PO_4)_{6-x}(OH)_{2-x} + (2-x)H^+ \rightarrow Ca_{10-x}(HPO_4)_x(PO_4)_{6-x}(H_2O)_{2-x}^{(2-x)+} \quad (1)$$

$$Ca_{10-x}(HPO_4)_x(PO_4)_{6-x}(H_2O)_{2-x}^{(2-x)+} \rightarrow 3Ca_3(PO_4)_2 + (1-x)Ca^{2+} + (2-x)H_2O \quad (2)$$

$$Ca_3(PO_4)_2 + 2H^+ \rightarrow Ca^{2+} + 2CaHPO_4 \quad (3)$$

$$CaHPO_4 + H^+ \rightarrow Ca^{2+} + H_2PO_4^- \quad (4)$$

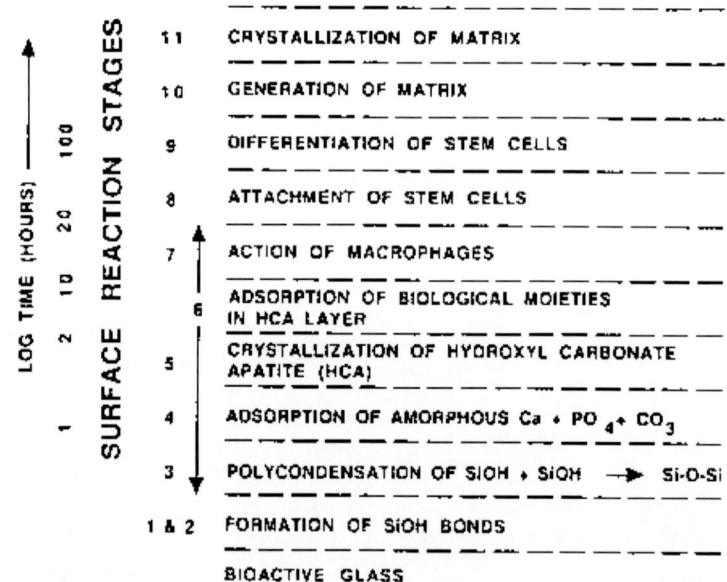

Figure 7. The sequence of interfacial reactions involved in forming a bond between tissue and bioactive glasses. The border between "dead" and "alive" occurs approximately at stage 6. For want of anything better, the bioactivity mechanism of calcium orthophosphates should also be described by this scheme with omitting of several initial stages, as it was made for HA in Ref. [72], where 3 initial chemical stages of the Hench's mechanism were replaced by partial dissolution of HA. Reprinted from Ref. [28] with permission.

Strange enough, but the bioactivity mechanism of calcium orthophosphates is not well described in literature; therefore, biomaterials researchers [72] are forced to use a modified scheme for the bioactivity mechanism of bioactive glasses – the concept introduced by Prof. Larry Hench [27,28]. The mechanism of bonding of bioactive glasses to living tissue involves a sequence of 11 successive reaction steps. The initial 5 steps occurred on the surface of bioactive glasses are "chemistry" only, while the remaining 6 steps belong to "biology" because the latter include colonization by osteoblasts, followed by proliferation and differentiation of the cells to form a new bone that had a mechanically strong bond to the implant surface (Fig. 7).

Biodegradability of polymers generally depends on the following factors: 1) chemical stability of the polymer backbone, 2) hydrophobicity of the monomer, 3) morphology of the polymer, 4) initial molecular weight, 5) fabrication processes, 6) geometry of the implant, 7) properties of the scaffold such as porosity and pore diameter [265]. A summary on degradation of PLA and PGA, as well as that of starch/ethylene vinyl alcohol copolymer (SEVA) is available in literature [Ref. 147, p. 798 and p. 803, respectively], where the interested readers are referred to. Biodegradation of HA/PLLA and CDHA/PLLA composite rods in subcutis and medullary cavities of rabbits were investigated mechanically and histologically; the degradation was found to be faster for the case of using uncalcinated CDHA instead of calcinated HA [934]. In a more detailed study, new bone formation was detected at 2 weeks after implantation, especially for formulations with a high HA content [935]. More to the point, a direct contact between bones and these composites without intervening fibrous tissue was detected in this case [935,936]. SEVA-C and SEVA-C/HA biocomposites were found to exhibit a non-cytotoxic behavior [937,938], inducing a satisfactory tissue response when implanted as shown by *in vivo* studies [938]. Furthermore, SEVA-C/HA biocomposites induce a positive response on osteoblast-like cells to what concerns cell adhesion and proliferation [937].

Both *in vitro* (the samples were immersed into 1% trypsin/phosphate-buffered saline solution at 37°C) and *in vivo* (implantation of samples into the posterolateral lumbar spine of rabbits) biodegradation have been investigated for nano-HA/collagen/PLA biocomposites [512]. The results demonstrated that weight loss increased continuously *in vitro* with a reduction in mass of 19.6% after 4 weeks. During the experimental period *in vitro*, the relative rate of reduction of the three components in this material was shown to differ greatly: collagen decreased the fastest, from 40% by weight to 20% in the composite; HA content increased from 45 to 60%; while PLA changed little. *In vivo*, the collagen/HA ratio appeared to be slightly higher near the transverse process than in the central part of the intertransverse process [512]. These data clearly demonstrate a biodegradation independence of various components of biocomposites.

Chapter 7

SOME CHALLENGES AND CRITICAL ISSUES

The scientific information summarized in this book represents the recent developments of calcium orthophosphate-based biocomposites and hybrid biomaterials from a variety of approaches, starting from conventional ones to tissue engineering. Such formulations combined with osteoconductive, osteoinductive factors, and/or osteogenic cells have gained much interest as a new and versatile class of biomaterials and are perceived to be beneficial in many aspects as bone grafts [32]. However, current applications of these biomaterials in medicine and surgery are still remarkably less than might be expected. In many biomedical applications, research and testing of such formulations have been introduced and highly developed but only in a very few cases an industrial production and commercial distribution of medical devices partially or entirely made of biocomposites have started. The medical application of biocomposites and hybrid biomaterials requires a better understanding of the objectives and limitations involved. Recently, the main critical issues have been summarized as follows [214]:

- There are not enough reliable experimental and clinical data supporting the long-term performance of biocomposites with respect to monolithic traditional materials.
- The design of biocomposites and hybrid biomaterials is far more complex than that of conventional monolithic materials because of the large number of additional design variables that must be considered.
- The available fabrication methods may limit the possible reinforcement configurations, may be time consuming, expensive,

- highly skilled and may require special cleaning and sterilization processes.
- There are no satisfactory standards yet for biocompatibility testing of the biocomposite implants because the ways in which the different components of any biocomposite interact to living tissues are not completely understood.
- There are no adequate standards for the assessment of biocomposite fatigue performance because the fatigue behavior of such materials is far more complex and difficult to predict than that of traditional materials [214].

On the other hand, in spite of an enormous progress in biocomposite processing, to achieve the desired characteristics researchers still need to develop more advanced technologies to fabricate a bone-resembling hierarchical organization over several length scales. Development of novel bone repair materials depends on the progress in research into the structure of natural bones. The key issues are not only to understand the fundamentals of biomineralization but also to translate such knowledge into practical synthetic pathways to produce better bone grafts. Unfortunately, when it comes to the fabrication of composites mimicing natural bone from the nanometer to the micrometer dimensions, there are many key issues, including control of morphology, incorporation of foreign ions, interaction with biomolecules and assembly of the organic and inorganic phases, which are still not well understood. A processing gap between the lower-level building units and the higher-order architecture could severely limit the practical application of current calcium orthophosphate-based biocomposites and hybrid biomaterials. Therefore, further substantial research efforts have been outlined to address the following key challenges [32,37]:

- Optimizing biocomposite processing conditions.
- Optimization of interfacial bonding and strength equivalent to natural bone.
- Optimization of the surface properties and pore size to maximize bone growth.
- Maintaining the adequate volume of the construct *in vivo* to allow bone formation to take place.
- Withstanding the load-bearing conditions.
- Matching the bioresorbability of the grafts and their biomechanical properties while forming new bone.

- Understanding the molecular mechanisms by which the cells and the biocomposite matrix interact with each other *in vivo* to promote bone regeneration.
- Supporting angiogenesis and vascularization for the growth of healthy bone cells and subsequent tissue formation and remodeling [32,37].

The aforementioned critical issues have to be solved before a widespread commercial use of calcium orthophosphate-based biocomposites and hybrid biomaterials can be made in surgery and medicine.

Chapter 8

CONCLUSIONS

All types of calcified tissues of humans and mammals appear to possess a complex hierarchical composite structure. Their mechanical properties are outstanding (considering weak constituents from which they are assembled) and far beyond those, that can be achieved using the same synthetic materials with present technologies. This is because biological organisms produce biocomposites that are organized in terms of both composition and structure, containing both brittle calcium orthophosphates and ductile bioorganic components in very complex structures, hierarchically organized at the nano-, micro- and meso-levels. Additionally, the calcified tissues are always multifunctional: for example, bone provides structural support for the body plus blood cell formation. The third defining characteristic of biological systems, in contrast with current synthetic systems, is their self-healing ability, which is nearly universal in nature. These complex structures, which have risen from millions of years of evolution, inspire materials scientists in the design of novel biomaterials [939].

Up to now, still no reasonable alternative exists to autogenous bone grafts in surgery. However, the studies summarized in this book have shown that the proper combination of a ductile matrix with a brittle, hard and bioactive calcium orthophosphate filler offers many advantages for biomedical applications. Namely, the desirable properties of some components can compensate for a poor mechanical behavior of calcium orthophosphate bioceramics, while in turn the desirable bioactive properties of calcium orthophosphates improve those of other phases, thus expanding the possible application of each material within the body [94]. However, the reviewed literature clearly indicates that among possible types of calcium orthophosphate-based biocomposites and hybrid biomaterials only simple, complex and graded ones (see classification of the composites in chapter 2 of this book) have been investigated. Presumably, a future progress in this subject will require concentrating efforts on elaboration and development of hierarchical biocomposites. Certainly, in the nearest future, much attention will be paid to further development of nano-sized and nanocrystalline calcium orthophosphates [940]. Furthermore, following the modern tendency of tissue engineering, a novel generation of calcium orthophosphate-based biocomposites and hybrid biomaterials should also contain a biological living part.

Much work remains to be done on a long way from a laboratory to clinics and the success in this field depends on the effective co-operation of clinicians, chemists, biologists, bioengineers and materials scientists.

Chapter 9

REFERENCES

[1] Keating JF, McQueen MM. Substitutes for autologous bone graft in orthopaedic trauma. *J. Bone Joint Surg.* (Br.) 2001;82B:3-8.
[2] Meyer U, Joos U, Wiesmann HP. Biological and biophysical principles in extracorporal bone tissue engineering: Part III. *Int. J. Oral Maxillofac. Surg.* 2004;33:635-641.
[3] Lane JM, Tomin E, Bostrom MPG. Biosynthetic bone grafting. *Clin. Orthop. Rel. Res.* 1999;367S:107-117.
[4] Murugan R, Ramakrishna S. Bioactive nanomaterials in bone grafting and tissue engineering. in: *Handbook of nanostructured biomaterials and their applications in nanobiotechnology.* Nalwa HS. (Ed.) American Scientific Publishers: Stevenson Ranch, 2005; Vol. 2, pp. 141-148.
[5] Keller EE, Triplett WW. Iliac crest bone grafting: review of 160 consecutive cases. *J. Oral Maxillofac. Surg.* 1987;45:11-14.
[6] Laurie SW, Kaban LB, Mulliken JB, Murray JE. Donor-site morbidity after harvesting rib and iliac bone. *Plast. Reconstr. Surg.* 1984;73:933-938.
[7] Younger EM, Chapman MW. Morbidity at bone graft sites. *J. Orthop. Trauma* 1989;3:192-195.
[8] Neumann M, Epple M. Composites of calcium phosphate and polymers as bone substitution materials. *Eur. J. Trauma* 2006;32:125-131.
[9] Le Guéhennec L, Layrolle P, Daculsi G. A review of bioceramics and fibrin sealant. *Eur. Cells Mater.* 2004;8:1-11.
[10] Fuchs JR, Nasseri BA, Vacanti JP. Tissue engineering: a 21st century solution to surgical reconstruction. Ann. Thorac. Surg. 2001;72:557-591.

[11] Hench LL, Wilson J. Surface-active biomaterials. *Science* 1984;226:630-636.
[12] Rose FRAJ, Oreffo ROC. Breakthroughs and views bone tissue engineering: hope vs. hype. *Biochem. Biophys. Res.* 2002;292:1-7.
[13] Kokubo T, Kim HM, Kawashita M. Novel bioactive materials with different mechanical properties. *Biomaterials* 2003;24:2161-2175.
[14] Rueger JM. Bone replacement materials – state of the art and the way ahead. *Orthopäde* 1998;27:72-79.
[15] Greenwald AS, Boden SD, Goldberg VM, Khan Y, Laurencin CT, Rosier RN. Bone graft substitutes: facts, fictions and applications. *J. Bone Joint Surg.* (Am.) 2001;83:98-103.
[16] Finkemeier CG. Bone grafting and bone graft substitutes. *J. Bone Joint Surg.* (Am.) 2002;84:454-464.
[17] Giannoudis PV, Dinopoulos H, Tsiridis E. *Bone substitutes: an update. Injury* 2005;36(Suppl. 3):S20-S27.
[18] Yang S, Leong KF, Du Z, Chua CK. The design of scaffolds for use in tissue engineering. Part I. Traditional factors. *Tissue Eng.* 2001;7:679-689.
[19] Burg KJL, Porter S, Kellam JF. Biomaterial developments for bone tissue engineering. *Biomaterials* 2000;21:2347-2359.
[20] Holy CE, Shoichet MS, Davies JE. Engineering three-dimensional bone tissue *in vitro* using biodegradable scaffolds: investigating initial cell-seeding density and culture period. *J. Biomed. Mater. Res.* 2000;51:376-382.
[21] Lowenstam HA, Weiner S. *On biomineralization.* New York: Oxford University Press, 1989.
[22] Weiner S, Wagner HD. The material bone: structure-mechanical function relations. *Ann. Rev. Mater. Sci.* 1998;28:271-298.
[23] Dorozhkin SV. Calcium orthophosphates in nature, biology and medicine. *Materials* 2009;2:399-498.
[24] Hench LL, Wilson J. in: An introduction to bioceramics. Hench LL, Wilson J. (Eds.). *Advanced series in ceramics, Vol 1.* World Scientific, Singapore, 1993, p. 1.
[25] Tadic D, Epple M. A thorough physicochemical characterisation of 14 calcium phosphate-based bone substitution materials in comparison to natural bone. *Biomaterials* 2004;25:987-994.
[26] Suchanek W, Yoshimura M. Processing and properties of hydroxyapatite-based biomaterials for use as hard tissue replacement implants. *J. Mater. Res.* 1998;13:94-117.

[27] Hench LL. Bioceramics: from a concept to clinics. *J. Am. Ceram. Soc.* 1991;74:1487-1510.
[28] Hench LL. Bioceramics. *J. Am. Ceram. Soc.* 1998;81:1705-1728.
[29] Itoh S, Kikuchi M, Koyama Y, Takakuda K, Shinomiya K, Tanaka J. Development of an artificial vertebral body using a novel biomaterial, hydroxyapatite/collagen composite. *Biomaterials* 2002;23:3919-3926.
[30] Thompson JB, Kindt JH, Drake B, Hansma HG, Morse DE, Hansma PK. Bone indentation recovery time correlates with bond reforming time. *Nature* 2001;414:773-776.
[31] Fratzl P, Gupta HS, Paschalis EP, Roschger P. Structure and mechanical quality of the collagen-mineral nano-composite in bone. *J. Mater. Chem.* 2004;14:2115-2123.
[32] Murugan R, Ramakrishna S. Development of nanocomposites for bone grafting. *Compos. Sci. Technol.* 2005;65:2385-2406.
[33] Burr DB. The contribution of the organic matrix to bone's material properties. *Bone* 2002;31:8-11.
[34] Itoh S, Kikuchi M, Koyama Y, Matumoto HN, Takakuda K, Shinomiya K, Tanaka J. Development of a novel biomaterial, hydroxyapatite/collagen composite for medical use. *Biomed. Mater. Eng.* 2005;15:29-41.
[35] Cui FZ, Li Y, Ge J. Self-assembly of mineralized collagen composites. *Mater. Sci. Eng.* R 2007;57:1-27.
[36] Vallet-Regi M, Arcos D. Nanostructured hybrid materials for bone tissue regeneration. *Current Nanoscietice* 2006;2:179-189.
[37] Chan CK, Kumar TSS, Liao S, Murugan R, Ngiam M, Ramakrishna S. Biomimetic nanocomposites for bone graft applications. *Nanomedicine* 2006;1:177-188.
[38] Olszta MJ, Cheng XG, Jee SS, Kumar BR, Kim YY, Kaufman MJ, Douglas EP, Gower LB. Bone structure and formation: A new perspective. *Mater. Sci. Eng.* R 2007;58:77-116.
[39] Bauer T, Muschler G. Bone grafts materials. An overview of the basic science. *Clin. Orthop. Rel. Res.* 2000;371:10-27.
[40] Athanasiou KA, Zhu CF, Lanctot DR, Agrawal CM, Wang X. Fundamentals of biomechanics in tissue engineering of bone. *Tissue Engineering* 2000;6:361-381.
[41] Zioupos P. Recent developments in the study of solid biomaterials and bone: "fracture" and "pre-fracture" toughness. *Mater. Sci. Eng. C* 1998;6:33-40.

[42] Doblaré M, Garcia JM, Gómez MJ. Modelling bone tissue fracture and healing: a review. *Eng. Fract. Mech.* 2004;71:1809-1840.
[43] Vallet-Regi M. *Revisiting ceramics for medical applications.* Dalton Trans. 2006;5211-5220.
[44] Huiskes R, Ruimerman R, Harry van Lenthe G, Janssen JD. Effects of mechanical forces on maintenance and adaptation of form in trabecular bone. *Nature* 2000;405:704-706.
[45] Thomson RC, Yaszemski MJ, Powers JM, Mikos AG. Hydroxyapatite fiber reinforced poly(α-hydroxy ester) foams for bone regeneration. *Biomaterials* 1998;19:1935-1943.
[46] Boccaccini AR, Blaker JJ. Bioactive composite materials for tissue engineering scaffolds. *Expert Rev. Med. Devices* 2005;2:303-317.
[47] Verheyen CCPM, de Wijn JR, van Blitterswijk CA, de Groot K, Rozing PM. Hydroxyapatite/poly(L-lactide) composites: an animal study on push-out strengths and interface histology. *J. Biomed. Mater. Res.* 1993;27:433-444.
[48] Zhang RY, Ma PX. Poly(α-hydroxyl acids) hydroxyapatite porous composites for bone-tissue engineering. I. Preparation and morphology. *J. Biomed. Mater. Res.* 1999;44:446-455.
[49] Durucan C, Brown PW. Biodegradable hydroxyapatite-polymer composites. *Adv. Eng. Mater.* 2001;3:227-231.
[50] Kim HW, Knowles JC, Kim HE. Hydroxyapatite/poly(ε-caprolactone) composite coatings on hydroxyapatite porous bone scaffold for drug delivery. *Biomaterials* 2004;25:1279-1287.
[51] Hutmacher DW, Schantz JT, Lam CXF, Tan KC, Lim TC. State of the art and future directions of scaffold-based bone engineering from a biomaterials perspective. *J. Tissue Eng. Regen. Med.* 2007;1:245-260.
[52] Guarino V, Causa F, Ambrosio L. Bioactive scaffolds for bone and ligament tissue. *Expert Review of Medical Devices* 2007;4:405-418.
[53] Yunos DM, Bretcanu O, Boccaccini AR. Polymer-bioceramic composites for tissue engineering scaffolds. *J. Mater. Sci.* 2008;43:4433-4442.
[54] Hench LL, Polak JM. Third-generation biomedical materials. *Science* 2002;295:1014-1017.
[55] Crane GM, Ishaug SL, Mikos AG. Bone tissue engineering. *Nature Medicine* 1995;1:1322-1324.
[56] LeGeros RZ. Calcium phosphate materials in restorative dentistry: a review. *Adv. Dent. Res.* 1988;2:164-180.

[57] Mathijsen A. *Nieuwe Wijze van Aanwending van het Gips-Verband bij Beenbreuken. Haarlem: J.B. van Loghem*; 1852.
[58] Dreesman H. Über Knochenplombierung. *Beitr. Klin. Chir.* 1892;9:804-810.
[59] Wang M. Developing bioactive composite materials for tissue replacement. *Biomaterials* 2003;24:2133-2151.
[60] http://en.wikipedia.org/wiki/Composite_material (assessed in May 2009).
[61] Evans SL, Gregson PJ. Composite technology in load-bearing orthopaedic implants. *Biomaterials* 1998;19:1329-1342.
[62] Habibovic P, Barrère F, van Blitterswijk CA, de Groot K, Layrolle P. Biomimetic hydroxyapatite coating on metal implants. *J. Am. Ceram. Soc.* 2002;85:517-522.
[63] Zhang RY, Ma PX. Biomimetic polymer/apatite composite scaffolds for mineralized tissue engineering. *Macromol. Biosci.* 2004;4:100-111.
[64] Oliveira AL, Mano JF, Reis RL. Nature-inspired calcium phosphate coatings: present status and novel advances in the science of mimicry. *Current Opinion in Solid State and Materials Science* 2003;7:309-318.
[65] Wan YZ, Hong L, Jia SR, Huang Y, Zhu Y, Wang YL, Jiang HJ. Synthesis and characterization of hydroxyapatite-bacterial cellulose nanocomposites. *Compos. Sci. Technol.* 2006;66:1825-1832.
[66] Wan YZ, Huang Y, Yuan CD, Raman S, Zhu Y, Jiang HJ, He F, Gao C. Biomimetic synthesis of hydroxyapatite/bacterial cellulose nanocomposites for biomedical applications. *Mater. Sci. Eng.* C 2007;27:855-864.
[67] Ohtsuki C, Kamitakahara M, Miyazaki T. Coating bone-like apatite onto organic substrates using solutions mimicking body fluid. *J. Tissue Eng. Regen. Med.* 2007;1:33-38.
[68] de Groot K, Geesink RGT, Klein CPAT, Serekian P. Plasma sprayed coatings of hydroxylapatite. *J. Biomed. Mater. Res.* 1987;21:1375-1381.
[69] de Groot K, Wolke JGC, Jansen JA. Calcium phosphate coatings for medical implants. *Proc. Instn. Mech. Engrs.* Part H 1998;212:137-147.
[70] Ignjatovic NL, Liu CZ, Czernuszka JT, Uskokovic DP. Micro- and nano-injectable composite biomaterials containing calcium phosphate coated with poly(DL-lactide-*co*-glycolide). *Acta Biomater.* 2007;3:927-935.
[71] Manso M, Langlet M, Fernandez M, Vasquez L, Martinez-Duart JM. Surface and interface analysis of hydroxyapatite/TiO_2 biocompatible structures. *Mater. Sci. Eng. C* 2003;23:451-454.

[72] Sun L, Berndt CC, Gross KA, Kucuk A. Material fundamentals and clinical performance of plasma-sprayed hydroxyapatite coatings: a review. J. Biomed. *Mater. Res. Appl. Biomater.* 2001;58:570-592.
[73] Yoshida K, Hashimoto K, Toda Y, Udagawa S, Kanazawa T. Fabrication of structure-controlled hydroxyapatite/zirconia composite. *J. Eur. Ceram. Soc.* 2006;26:515-518.
[74] Planeix JM, Jaunky W, Duhoo T, Czernuszka JT, Hosseini MW, Brès EF. A molecular tectonics-crystal engineering approach for building organic-inorganic composites. Potential application to the growth control of hydroxyapatite crystals. *J. Mater. Chem.* 2003;13:2521-2524.
[75] Dong J, Uemura T, Kojima H, Kikuchi M, Tanaka J, Tateishi T. Application of low-pressure system to sustain *in vivo* bone formation in osteoblastrporous hydroxyapatite composite. *Mater. Sci. Eng. C* 2001;17:37-43.
[76] Zerbo IR, Bronckers ALJJ, de Lange G, Burger EH. Localisation of osteogenic and osteoclastic cells in porous β-tricalcium phosphate particles used for human maxillary sinus floor elevation. *Biomaterials* 2005;26:1445-1451.
[77] Krout A, Wen HB, Hippensteel E, Li P. A hybrid coating of biomimetic apatite and osteocalcin. *J. Biomed. Mater. Res.* A 2005;73A:377-387.
[78] Matthews FL, Rawlings RD. *Composite materials: engineering and science.* CRC Press LLC, Boca Raton FL, 2000, 480 pp.
[79] Xia Z, Riester L, Curtin WA, Li H, Sheldon BW, Liang J, Chang B, Xu JM. Direct observation of toughening mechanisms in carbon nanotube ceramic matrix composites. *Acta Mater.* 2004;52:931-944.
[80] Williams DF. Composites. In: Williams DF. (Ed.). *Encyclopedia of Biomaterials.* Oxford, Pergamon, 1990.
[81] Ong JL, Chan DCN. Hydroxyapatite and their use as coatings in dental implants: a review. *Crit. Rev. Biomed. Eng.* 1999;28:667-707.
[82] Davies JE. *In vitro* modeling of the bone/implant interface. *Anat. Record* 1996;245:426-445.
[83] Anselme K. Osteoblast adhesion on biomaterials. *Biomaterials* 2000;21:667-681.
[84] Gauthier O, Bouler JM, Weiss P, Bosco J, Daculsi G, Aguado E. Kinetic study of bone ingrowth and ceramic resorption associated with the implantation of different injectable calcium-phosphate bone substitutes. *J. Biomed. Mater. Res.* 1999;47:28-35.
[85] Hing KA, Best SM, Bonfield W. Characterization of porous hydroxyapatite. *J. Mater. Sci. Mater. Med.* 1999;10:135-145.

[86] Carotenuto G, Spagnuolo G, Ambrosio L, Nicolais L. Macroporous hydroxyapatite as alloplastic material for dental applications. *J. Mater. Sci. Mater. Med.* 1999;10:671-676.
[87] LeGeros RZ. Calcium phosphates in oral biology and medicine. *Monographs in oral science.* Vol. 15. Myers HM. (Ed.), Karger, Basel, 1991, 201 pp.
[88] Elliot JC. Structure and chemistry of the apatites and other calcium orthophosphates. *Studies in inorganic chemistry.* Vol. 18. Elsevier, Amsterdam – London – New York – Tokyo, 1994, 389 pp.
[89] Brown PW, Constantz B. (Eds.) *Hydroxyapatite and related materials.* CRC Press, Boca Raton – Ann Arbor – London – Tokyo, 1994, 343 pp.
[90] Amjad Z. (Ed.) *Calcium phosphates in biological and industrial systems.* Kluwer Academic Publishers, Boston, MA, 1997, 529 pp.
[91] Hughes JM, Kohn M, Rakovan J. (Eds.) Phosphates: geochemical, geobiological and materials importance. Series: *Reviews in mineralogy and geochemistry. Vol. 48.* Mineralogical Society of America, Washington, DC, 2002.
[92] Chow LC, Eanes ED. (Eds.) Octacalcium phosphate. *Monographs in oral science. Vol. 18*, S. Karger AG, Basel, 2001, 168 pp.
[93] Brès E, Hardouin P. (Eds.) Les matériaux en phosphate de calcium. Aspects fondamentaux. / Calcium phosphate materials. *Fundamentals. Sauramps Medical, Montpellier,* 1998, 176 pp.
[94] Rea SM, Bonfield W. Biocomposites for medical applications. *J. Aust. Ceram. Soc.* 2004;40:43-57.
[95] Langer R. Biomaterials in drug delivery and tissue engineering: one laboratory's experience. *Acc. Chem. Res.* 2000;33:94-101.
[96] Thomson RC, Ak S, Yaszemski MJ, Mikos AG. *Polymer scaffold processing. In: Principles of Tissue Engineering.* Academic Press, NY, USA, 2000, pp. 251-262.
[97] Ramakrishna S, Mayer J, Wintermantel E, Leong KW. Biomedical applications of polymer-composite materials: a review. *Compos. Sci. Technol.* 2001;61:1189-1224.
[98] Langer R, Vacanti JP. Tissue engineering. *Science* 1993;260:920-925.
[99] Lanza RP, Hayes JL, Chick WL. Encapsulated cell technology. *Nature Biotechnology* 1996;14:1107-1111.
[100] Agrawal CM, Ray RB. Biodegradable polymeric scaffolds for musculoskeletal tissue engineering. *J. Biomed. Mater. Res.* 2001;55:141-150.

[101] Kweon H, Yoo M, Park I, Kim T, Lee H, Lee S, Oh J, Akaike T, Cho C. A novel degradable polycaprolactone network for tissue engineering. *Biomaterials* 2003;24:801-808.

[102] de Groot JH, de Vrijer R, Pennings AJ, Klompmaker J, Veth RPH, Jansen HWB. Use of porous polyurethanes for meniscal reconstruction and meniscal prostheses. *Biomaterials* 1996;17:163-173.

[103] Resiak I, Rokicki G. Modified polyurethanes for biomedical applications. *Polimery* 2000;45:592-602.

[104] Temenoff JS, Mikos AG. Injectable biodegradable materials for orthopedic tissue engineering. *Biomaterials* 2000;21:2405-2412.

[105] Behravesh E, Yasko AW, Engel PS, Mikos AG. Synthetic biodegradable polymers for orthopaedic applications. *Clin. Orthop. Rel. Res.* 1999;367S:S118-S125.

[106] Lewandrowski KU, Gresser JD, Wise DL, White RL, Trantolo DJ. Osteoconductivity of an injectable and bioresorbable poly(propyleneglycol-*co*-fumaric acid) bone cement. *Biomaterials* 2000;21:293-298.

[107] Peter SJ, Miller MJ, Yaszemski MJ, Mikos AG. Poly(propylene fumarate). In: Domb AJ, Kost J, Wiseman DM, editors. *Handbook of biodegradable polymers*. Amsterdam: Harwood Academic, 1997. pp. 87-97.

[108] Boland ED, Coleman BD, Barnes CP, Simpson DG, Wnek GE, Bowlin GL. Electrospinning polydioxanone for biomedical applications. *Acta Biomater.* 2005;1:115-123.

[109] Gilbert JL. A*crylics in biomedical engineering.* In: *Encyclopedia of materials: science and technology*. Elsevier, Amsterdam, 2001, pp. 11-18.

[110] Li YW, Leong JCY, Lu WW, Luk KDK, Cheung KMC, Chiu KY, Chow SP. A novel injectable bioactive bone cement for spinal surgery: a developmental and preclinical study. *J. Biomed. Mater. Res.* 2000;52:164-170.

[111] Mckellop H, Shen F, Lu B, Campbell P, Salovey R. Development of an extremely wear resistant UHMW polyethylene for total hip replacements. *J. Orthop. Res.* 1999;17:157-167.

[112] Kurtz SM, Muratoglu OK, Evans M, Edidin AA. Advances in the processing, sterilization and crosslinking of ultra-high molecular weight polyethylene for total joint arthroplasty. *Biomaterials* 1999;20:1659-1688.

[113] Laurencin CT, Ambrosio MA, Borden MD, Cooper JA, Jr. Tissue engineering: orthopedic applications. *Ann. Rev. Biomed. Eng.* 1999;1:19-46.
[114] Jansen JA, de Ruijter JE, Janssen PT, Paquay YG. Histological evaluation of biodegradable Polyactive™/hydroxyapatite membrane. *Biomaterials* 1995;16:819-827.
[115] Liu Q, de Wijn JR, Bakker D, van Blitterswijk CA. Surface modification of hydroxyapatite to introduce interfacial bonding with Polyactive™ 70/30 in a biodegradable composite. *J. Mater. Sci. Mater. Med.* 1996;7:551-557.
[116] Liu Q, de Wijn JR, Bakker D, van Toledo M, van Blitterswijk CA. Polyacids as bonding agents in hydroxyapatite polyesterether (Polyactive™ 30/70) composite. *J. Mater. Sci. Mater. Med.* 1998;9:23-30.
[117] Meijer GJ, Cune MS, van Dooren M, de Putter C, van Blitterswijk CA. A comparative study of flexible (Polyactive™) versus rigid (hydroxylapatite) permucosal dental implants. I. Clinical aspects. *J. Oral Rehabilitation* 1997;24:85-92.
[118] Meijer GJ, Dalmeijer RA, de Putter C, van Blitterswijk CA. A comparative study of flexible (Polyactive™) versus rigid (hydroxylapatite) permucosal dental implants. II. Histological aspects. *J. Oral Rehabilitation* 1997;24:93-101.
[119] Svensson A, Nicklasson E, Harrah T, Panilaitis B, Kaplan DL, Brittberg M, Gatenholm P. Bacterial cellulose as a potential scaffold for tissue engineering of cartilage. *Biomaterials* 2005;26:419-431.
[120] Granja PL, Barbosa MA, Pouysége L, de Jéso B, Rouais F, Baquuey C. Cellulose phosphates as biomaterials. Mineralization of chemically modified regenerated cellulose hydrogels. *J. Mater. Sci.* 2001;36:2163-2172.
[121] Thomas V, Dean DR, Vohra YK. Nanostructured biomaterials for regenerative medicine. *Current Nanoscience* 2006;2:155-177.
[122] Li SM, Garreau H, Vert M. Structure-property relationships in the case of the degradation of massive aliphatic poly(α-hydroxyacids) in aqueous media. Part 1: poly(DL-lactic acid). *J. Mater. Sci. Mater. Med.* 1990;1:123-130.
[123] Daniels AU, Adriano KP, Smuts WP, Chang MKO, Keller J. Evaluation of absorbable poly(orthoesters) for use in surgical implants. *J. Appl. Biomater.* 1994;5:51-64.

[124] Adriano KP, Pohjonen T, Törmällä P. Processing and characterization of absorbable polylactide polymers for use in surgical implants. *J. Appl. Biomater.* 1994;5:133-140.
[125] Athanasiou KA, Niederauer GG, Agrawal CM. Sterilization, toxicity, biocompatibility and clinical applications of polylactic acid/polyglycolic acid copolymers. *Biomaterials* 1996;17:93-102.
[126] Dee KC, Bizios R. Mini-review: proactive biomaterials and bone tissue engineering. *Biotechnology and Bioengineering* 1996;50:438-442.
[127] Ignjatovic N, Tomic S, Dakic M, Miljkovic M, Plavsic M, Uskokovic D. Synthesis and properties of hydroxyapatite/poly-L-lactide composite biomaterials. *Biomaterials* 1999;20:809-816.
[128] Ignjatovic N, Savic V, Najman S, Plavsic M, Uskokovic D. A study of HAp/PLLA composite as a substitute for bone powder using FT-IR spectroscopy. *Biomaterials* 2001;22:571-575.
[129] Marra KG, Szem JW, Kumta PN, DiMilla PA, Weiss LE. *In vitro* analysis of biodegradable polymer blend/hydroxyapatite composites for bone tissue engineering. *J. Biomed. Mater. Res.* 1999;47:324-335.
[130] Ashammakhi N, Rokkanen P. Absorbable polyglycolide devices in trauma and bone surgery. *Biomaterials* 1997;18:3-9.
[131] Boyan B, Lohmann C, Somers A, Neiderauer G, Wozney J, Dean D, Carnes D, Schwartz Z, Potential of porous poly-D,L-lactide-*co*-glycolide particles as a carrier for recombinant human bone morphogenetic protein-2 during osteoinduction *in vivo*. *J. Biomed. Mater. Res.* 1999;46:51-59.
[132] Hofmann GO. Biodegradable implants in traumatology: a review of the state-of-the-art. *Arch. Orthopedic Trauma Surgery* 1995;114:123-132.
[133] Hollinger JO, Leong K. Poly(α-hydroxyacids): carriers for bone morphogenetic proteins. *Biomaterials* 1996;17:187-194.
[134] Griffith LG. Polymeric biomaterials. *Acta Materialia* 2000;48:263-277.
[135] Peter SJ, Miller MJ, Yasko AW, Yaszemski MJ, Mikos AG. Polymer concepts in tissue engineering. *J. Biomed. Mater. Res.* 1998;43:422-427.
[136] Ishuang SL, Payne RG, Yaszemski MJ, Aufdemorte TB, Bizios R, Mikos AG. Osteoblast migration on poly(α-hydroxy esters). *Biothechnology and Bioengineering* 1996;50:443-451.
[137] Shikinami Y, Okuno M. Bioresorbable devices made of forged composites of hydroxyapatite (HA) particles and poly-L-lactide (PLLA): Part I. Basic characteristics. *Biomaterials* 1999;20:859-877.
[138] Khor E, Lim LY. Implantable applications of chitin and chitosan. *Biomaterials* 2003;24:2339-2349.

[139] Ishihara M, Nakanishi K, Ono K, Sato M, Kikuchi M. Photocrosslinkable chitosan as a dressing for wound occlusion and accelerator in healing process. *Biomaterials* 2002;23:833-840.

[140] Di Martino A, Sittinger M, Risbud MV. Chitosan: a versatile biopolymer for orthopaedic tissue-engineering. *Biomaterials* 2005;26:5983-5990.

[141] Piskin E, Bölgen N, Egri S, Isoglu IA. Electrospun matrices made of poly(α-hydroxy acids) for medical use. *Nanomedicine* 2007;2:441-457.

[142] Rezwana K, Chena QZ, Blakera JJ, Boccaccini AR. Biodegradable and bioactive porous polymer/inorganic composite scaffolds for bone tissue engineering. *Biomaterials* 2006;27:3413-3431.

[143] Mohanty AK, Misra M, Hinrichsen G. Biofibres, biodegradable polymers and biocomposites: an overview. *Macromol. Mater. Eng.* 2000;276/277:1-24.

[144] Kohane DS, Langer R. Polymeric biomaterials in tissue engineering. *Pediatric Res.* 2008;63:487-491.

[145] Seal BL, Otero TC, Panitch A. Polymeric biomaterials for tissue and organ regeneration. *Mater. Sci. Eng. R* 2001;34:147-230.

[146] An YH, Woolf SK, Friedman RJ. Pre-clinical *in vivo* evaluation of orthopaedic bioabsorbable devices. *Biomaterials* 2000;21:2635-2652.

[147] Mano JF, Sousa RA, Boesel LF, Neves NM, Reis RL. Bioinert, biodegradable and injectable polymeric matrix composites for hard tissue replacement: state of the art and recent developments. *Compos. Sci. Technol.* 2004;64:789-817.

[148] Katti KS. Biomaterials in total joint replacement. Colloids Surf. *B Biointerfaces* 2004;39:133-142.

[149] Hayashi T. Biodegradable polymers for biomedical uses. *Prog. Polym. Sci.* 1994;19:663-702.

[150] Middleton J, Tipton A. Synthetic biodegradable polymers as orthopedic devices. *Biomaterials* 2000;21:2335-2346.

[151] Ma PX. Biomimetic materials for tissue engineering. *Adv. Drug Delivery Rev.* 2008;60:184-198.

[152] Coombes AG, Meikle MC. Resorbable synthetic polymers as replacements for bone graft. *Clin. Mater.* 2004;17:35-67.

[153] Okada M. Chemical syntheses of biodegradable polymers. *Prog. Polym. Sci.* 2002;27:87-133.

[154] Jordan J, Jacob KI, Tannenbaum R, Sharaf MA, Jasiuk I. Experimental trends in polymer nanocomposites – a review. *Mater. Sci. Eng. A* 2005;393:1-11.

[155] Matsuno H, Yokoyama A, Watari F, Uo M, Kawasaki T. Biocompatibility and osteogenesis of refractory metal implants, titanium, hafnium, niobium, tantalum and rhenium. *Biomaterials* 2001;22:1253-1262.
[156] Uo M, Watari F, Yokoyama A, Matsuno H, Kawasaki T. Dissolution of nickel and tissue response observed by X-ray analytical microscopy. *Biomaterials* 1999;20:747-755.
[157] Uo M, Watari F, Yokoyama A, Matsuno H, Kawasaki T. Visualization and detectability of rarely contained elements in soft tissue by X-ray scanning analytical microscopy and electron probe micro analysis. *Biomaterials* 2001;22:1787-1794.
[158] Uo M, Watari F, Yokoyama A, Matsuno H, Kawasaki T. Tissue reaction around metal implants observed by X-ray scanning analytical microscopy. *Biomaterials* 2001;21:677-685.
[159] Ryan G, Pandit A, Apatsidis DP. Fabrication methods of porous metals for use in orthopaedic applications. *Biomaterials* 2006;27:2651-2670.
[160] Shimko DA, Shimko VF, Sander EA, Dickson KF, Nauman EA. Effect of porosity on the fluid flow characteristics and mechanical properties of tantalum scaffolds. *J. Biomed. Mater. Res. B Appl. Biomater.* 2005;73B:315-324.
[161] Wen CE, Yamada Y, Shimojima K, Chino Y, Hosokawa H, Mabuchi M. Compressibility of porous magnesium foam: dependency on porosity and pore size. *Mater. Lett.* 2004;58:357-360.
[162] Green D, Walsh D, Mann S, Oreffo ROC. The potentials of biomimesis in bone tissue engineering: lessons from the design and synthesis of invertebrate skeletons. *Bone* 2002;30:810-815.
[163] Kokubo T. Apatite formation on surfaces of ceramics, metals and polymers in body environment. *Acta Mater.* 1998;46:2519-2527.
[164] Witte F, Reifenrath J, Müller PP, Crostack HA, Nellesen J, Bach FW, Bormann D, Rudert M. Cartilage repair on magnesium scaffolds used as a subchondral bone replacement. *Mat.-wiss. u. Werkstofflech.* 2006;37:504-508.
[165] Uo M, Mizuno M, Kuboki Y, Makishima A, Watari F. Properties and cytotoxicity of water soluble Na_2O-CaO-P_2O_5 glasses. *Biomaterials* 1998;19:2277-2284.
[166] Imai T, Watari F, Yamagata S, Kobayashi M, Nagayama K, Nakamura S. Mechanical properties and estheticity of FRP orthodontic wire fabricated by hot drawing. *Biomaterial* 1998;19:2195-2200.

[167] Watari F, Yamagata S, Imai T, Nakamura S, Kobayashi M. The fabrication and properties of aesthetic FRP wires for use in orthodontics. *J. Mater. Sci.* 1998;33:5661-5664.
[168] Hench LL. The story of Bioglass®. *J. Mater. Sci. Mater. Med.* 2006;17:967-978.
[169] Cao W, Hench LL. Bioactive materials. *Ceramics International* 1996;22:493-507.
[170] Vogel M, Voigt C, Gross U, Müller-Mai C. In vivo comparison of bioactive glass particles in rabbits. *Biomaterials* 2001;22:357-362.
[171] de Aza P, Luklinska Z, Santos C, Guitian F, de Aza S. Mechanism of bone-like formation on a bioactive implant *in vivo*. *Biomaterials* 2003;24:1437-1445.
[172] Ikeda N, Kawanabe K, Nakamura T. Quantitative comparison of osteoconduction of porous, dense A-W glass-ceramic and hydroxyapatite granules (effects of granule and pore sizes). *Biomaterials* 1999;20:1087-1095.
[173] Weizhong Y, Dali Z, Guangfu Y. Research and development of A-W bioactive glass ceramic. *J. Biomed. Engin.* 2003;20:541-545.
[174] Piconi C, Maccauro G. Zirconia as a ceramic biomaterial. *Biomaterials* 1999;20:1-25.
[175] Christel P, Meunier A, Heller M, Torre JP, Peille CN. Mechanical properties and short-term *in vivo* evaluation of yttrium-oxide-partially-stabilized-zirconia. *J. Biomed. Mater. Res.*1989;23:45-61.
[176] Garvie RC, Urban D, Kennedy DR, McMeuer JC. Biocompatibility of magnesium partially stabilized zirconia (Mg-PSZ ceramics). *J. Mater. Sci.* 1984;19:3224-3228.
[177] Burger W, Richter HG, Piconi C, Vatteroni R, Cittadini A, Bocccalari M. New Y-TZP powders for medical grade zirconia. *J. Mater. Sci. Mater. Med.* 1997;8:113-118.
[178] Converse GL, Yue W, Roeder RK. Processing and tensile properties of hydroxyapatite-whisker-reinforced polyetheretherketone. *Biomaterials* 2007;28:927-935.
[179] Yue W, Roeder RK. Micromechanical model for hydroxyapatite whisker reinforced polymer biocomposites. *J. Mater. Res.* 2006;21:2136-2145.
[180] Converse GL, Roeder RK. Tensile properties of hydroxyapatite whisker reinforced polyetheretherketone. *Mater. Res. Soc. Symp. Proc.* 2005;898:44-49.
[181] Mizutani Y, Hattori M, Okuyama M, Kasuga T, Nogami M. preparation of porous composites with a porous framework using hydroxyapatite

whiskers and poly(L-lactic acid) short fibers. *Key Engin. Mater.* 2006;309-311:1079-1082.

[182] Watanabe T, Ban S, Ito T, Tsuruta S, Kawai T, Nakamura H. Biocompatibility of composite membrane consisting of oriented needle-like apatite and biodegradable copolymer with soft and hard tissues in rats. *Dental Materials J.* 2004;23:609-612.

[183] Li H, Chen Y, Xie Y. Photo-crosslinking polymerization to prepare polyanhydride/needle-like hydroxyapatite biodegradable nanocomposite for orthopedic application. *Mater. Lett.* 2003;57:2848-2854.

[184] Peng Q, Weng J, Li X, Gu Z. Manufacturing porous blocks of nano-composite of needle-like hydroxyapatite crystallites and chitin for tissue engineering. *Key Engin. Mater.* 2005;288-289:199-202.

[185] Kasuga T, Ota Y, Nogami M, Abe Y. Preparation and mechanical properties of polylactic acid composites containing hydroxyapatite fibers. *Biomaterials* 2000;22:19-23.

[186] Smith L. Ceramic-plastic material as a bone substitute. *Arch. Surg.* 1963;87:653-661.

[187] Bonfield W, Grynpas MD, Tully AE, Bowman J, Abram J. Hydroxyapatite reinforced polyethylene – a mechanically compatible implant material for bone replacement. *Biomaterials* 1981;2:185-189.

[188] Bonfield W, Bowman J, Grynpas MD. *Composite material for use in orthopaedics*. UK Patent 8032647, 1981.

[189] Bonfield W. Composites for bone replacement. *J. Biomed. Eng.* 1988;10:522-526.

[190] Wang M, Porter D, Bonfield W. Processing, characterisation and evaluation of hydroxyapatite reinforced polyethylene composites. *British Ceram. Trans.* 1994;93:91-95.

[191] Guild FJ, Bonfield W. Predictive character of hydroxyapatite-polyethelene HAPEX™ composite. *Biomaterials* 1993;14:985-993.

[192] Huang J, Di Silvio L, Wang M, Tanner KE, Bonfield W. *In vitro* mechanical and biological assessment of hydroxyapatite-reinforced polyethylene composite. *J. Mater. Sci. Mater. Med.* 1997;8:775-779.

[193] Deb S, Wang M, Tanner KE, Bonfield W. Hydroxyapatite-polyethylene composites: effect of grafting and surface treatment of hydroxyapatite. *J. Mater. Sci. Mater. Med.* 1996;7:191-193.

[194] Wang M, Joseph R, Bonfield W. Hydroxyapatite-polyethylene composites for bone substitution: effect of ceramic particle size and morphology. *Biomaterials* 1998;19:2357-2366.

[195] Suwanprateeb J, Tanner KE, Turner S, Bonfield W. Influence of Ringer's solution on creep resistance of hydroxyapatite reinforced polyethylene composites. *J. Mater. Sci. Mater. Med.* 1997;8:469-472.
[196] Ladizesky NH, Ward IM, Bonfield W. Hydroxyapatite/high-performance polyethylene fiber composites for high load bearing bone replacement materials. *J. Appl. Polym. Sci.* 1997;65:1865-1882.
[197] Ladizesky NH, Pirhonen EM, Appleyard DB, Ward IM, Bonfield W. Fibre reinforcement of ceramic/polymer composites for a major load-bearing bone substitute material. *Compos. Sci. Technol.* 1998;58:419-434.
[198] Nazhat SN, Joseph R, Wang M, Smith R, Tanner KE, Bonfield W. Dynamic mechanical characterisation of hydroxyapatite reinforced polyethylene: effect of particle size. *J. Mater. Sci. Mater. Med.* 2000;11:621-628.
[199] Ladizesky NH, Ward IM, Bonfield W. Hydrostatic extrusion of polyethylene filled with hydroxyapatite. *Polym. Adv. Technol.* 1996;8:496-504.
[200] Guild FJ, Bonfield W. Predictive modelling of the mechanical properties and failure processes of hydroxyapatite-polyethylene (HAPEX™) composite. *J. Mater. Sci. Mater. Med.* 1998;9:497-502.
[201] Di Silvio L, Dalby M, Bonfield W. *In vitro* response of osteoblasts to hydroxyapatite/reinforced polyethylene composites. *J. Mater. Sci. Mater. Med.* 1998;9:845-848.
[202] Wang M, Ladizesky NH, Tanner KE, Ward IM, Bonfield W. Hydrostatically extruded HAPEX™. *J. Mater. Sci.* 2000;35:1023-1030.
[203] That PT, Tanner KE, Bonfield W. Fatigue characterization of a hydroxyapatite-reinforced polyethylene composite. I. Uniaxial fatigue. *J. Biomed. Mater. Res.* 2000;51:453-460.
[204] That PT, Tanner KE, Bonfield W. Fatigue characterization of a hydroxyapatite-reinforced polyethylene composite. II. Biaxial fatigue. *J. Biomed. Mater. Res.* 2000;51:461-468.
[205] Bonner M, Ward IM, McGregor W, Tanner KE, Bonfield W. Hydroxyapatite/polypropylene composite: A novel bone substitute material. *J. Mater. Sci. Lett.* 2001;20:2049-2052.
[206] Bonner M, Saunders LS, Ward IM, Davies GW, Wang M, Tanner KE, Bonfield W. Anisotropic mechanical properties of oriented HAPEX™. *J. Mater. Sci.* 2002;37:325-334.

[207] Dalby MJ, Di Silvio L, Davies GW, Bonfield W. Surface topography and HA filler volume effect on primary human osteoblasts *in vitro*. *J. Mater. Sci. Mater. Med.* 2000;12:805-810.
[208] Di Silvio L, Dalby MJ, Bonfield W. Osteoblast behaviour on HA/PE composite surfaces with different HA volumes. *Biomaterials* 2002;23:101-107.
[209] Dalby MJ, Kayser MV, Bonfield W, Di Silvio L. Initial attachment of osteoblasts to an optimised HAPEX™ topography. *Biomaterials* 2002;23:681-690.
[210] Dalby MJ, Di Silvio L, Gurav N, Annaz B, Kayser MV, Bonfield W. Optimizing HAPEX™ topography influences osteoblast response. *Tissue Eng.* 2002;8:453-467.
[211] Zhang Y, Tanner KE, Gurav N, Di Silvio L. *In vitro* osteoblastic response to 30 vol% hydroxyapatite-polyethylene composite. *J. Biomed. Mater. Res. A* 2007;81A:409-417.
[212] Rea SM, Best SM, Bonfield W. Bioactivity of ceramic-polymer composites with varied composition and surface topography. *J. Mater. Sci. Mater. Med.* 2004;15:997-1005.
[213] Rea SM, Brooks RA, Schneider A, Best SM, Bonfield W. Osteoblast-like cell response to bioactive composites-surface-topography and composition effects. *J Biomed Mater Res B Appl Biomater.* 2004;70:250-261.
[214] Salernitano E, Migliaresi C. Composite materials for biomedical applications: a review. *J. Appl. Biomaterials & Biomechanics* 2003;1:3-18.
[215] Pandey A, Jan E, Aswath PB. Physical and mechanical behavior of hot rolled HDPE/HA composites. *J. Mater. Sci.* 2006;41:3369-3376.
[216] Sousa RA, Reis RL, Cunha AM, Bevis MJ. Processing and properties of bone-analogue biodegradable and bioinert polymeric composites. *Compos. Sci. Technol.* 2003;63:389-402.
[217] Homaeigohar SS, Shokrgozar MA, Khavandi A, Sadi AY. *In vitro* biological evaluation of β-TCP/HDPE – a novel orthopedic composite: A survey using human osteoblast and fibroblast bone cells. *J. Biomed. Mater. Res. A* 2008;84A:491-499.
[218] Downes RN, Vardy S, Tanner KE, Bonfield W. Hydroxyapatite-polyethylene composite in orbital surgery. *Bioceramics* 4. 1991, pp. 239-246.
[219] Dornhoffer HL. Hearing results with the dornhoffer ossicular replacement prostheses. *Laryngoscope* 1998;108:531-536.

[220] Swain RE, Wang M, Beale B, Bonfield W. HAPEX™ for otologic applications. *Biomed. Eng. Appl. Basis Commun.* 1999;11:315-320.
[221] Yi Z, Li Y, Jidong L, Xiang Z, Hongbing L, Yuanyuan W, Weihu Y. Novel bio-composite of hydroxyapatite reinforced polyamide and polyethylene: composition and properties. *Mater. Sci. Eng. A* 2007;452-453:512-517.
[222] Unwin AP, Ward IM, Ukleja P, Weng J. The role of pressure annealing in improving the stiffness of polyethylene/hydroxyapatite composites. *J. Mater. Sci.* 2001;36:3165-3177.
[223] Fang LM, Leng Y, Gao P. Processing and mechanical properties of HA/UHMWPE nanocomposites. *Biomaterials* 2006;27:3701-3707.
[224] Fang LM, Gao P, Leng Y. High strength and bioactive hydroxyapatite nano-particles reinforced ultrahigh molecular weight polyethylene. *Composites B* 2007;38:345-351.
[225] Fang LM, Leng Y, Gao P. Processing of hydroxyapatite reinforced ultrahigh molecular weight polyethylene for biomedical applications. *Biomaterials* 2005;26:3471-3478.
[226] Sousa RA, Reis RL, Cunha AM, Bevis MJ. Structure development and interfacial interactions in high-density polyethylene/hydroxyapatite (HDPE/HA) composites molded with preferred orientation. *J. Appl. Polym. Sci.* 2002;86:2873-2886.
[227] Reis RL, Cunha AM, Oliveira MJ, Campos AR, Bevis MJ. Relationship between processing and mechanical properties of injection molded high molecular mass polyethylene + hydroxyapatite composites. *Mat. Res. Innovat.* 2001;4:263-272.
[228] Donners JJJM, Nolte RJM, Sommerdijk NAJM. Dendrimer-based hydroxyapatite composites with remarkable materials properties. *Adv. Mater.* 2003;15:313-316.
[229] Ignjatovic NL, Plavsic M, Miljkovic MS, Zivkovic LM, Uskokovic DP. Microstructural characteristics of calcium hydroxyapatite/poly-L-lactide based composites. *J. Microscopy Oxford* 1999;196:243-248.
[230] Skrtic D, Antonucci JM, Eanes ED. Amorphous calcium phosphate-based bioactive polymeric composites for mineralized tissue regeneration. *J. Res. Natl. Inst. Stand. Technol.* 2003;108:167-182.
[231] Rizzi SC, Heath DJ, Coombes AGA, Bock N, Textor M, Downes S. Biodegradable polymer/hydroxyapatite composites: surface analysis and initial attachment of human osteoblasts. *J. Biomed. Mater. Res.* 2001;55:475-486.

[232] Kato K, Eika Y, Ikada Y. *In situ* hydroxyapatite crystallization for the formation of hydroxyapatite/polymer composites. *J. Mater. Sci.* 1997;32:5533-5543.
[233] Damien CJ, Parsons JR. Bone-graft and bone-graft substitutes a review of current technology and applications. *J. Appl. Biomater.* 1991;2:187-208.
[234] Zhang RY, Ma PX. Porous poly(L-lactic acid)/apatite composites created by biomimetic process. *J. Biomed. Mater. Res.* 1999;45:285-293.
[235] Liu Q, de Wijn JR, van Blitterswijk CA. Composite biomaterials with chemical bonding between hydroxyapatite filler particles and PEG/PBT copolymer matrix. *J. Biomed. Mater. Res.* 1998;40:490-497.
[236] Cerrai P, Guerra GD, Tricoli M, Krajewski A, Ravaglioli A, Martinetti R, Dolcini L. Fini M, Scarano A, Piattelli A. Periodontal membranes from composites of hydroxyapatite and bioresorbable block copolymers. *J. Mater. Sci. Mater. Med.* 1999;10:677-682.
[237] Roeder RK, Sproul MM, Turner CH. Hydroxyapatite whiskers provide improved mechanical properties in reinforced polymer composites. *J. Biomed. Mater. Res. A* 2003;67A:801-812.
[238] Hutmacher DW. Scaffolds in tissue engineering bone and cartilage. *Biomaterials* 2000;21:2529-2543.
[239] Mathieu LM, Bourban PE, Manson JAE. Processing of homogeneous ceramic/polymer blends for bioresorbable composites. *Compos. Sci. Technol.* 2006;66:1606-1614.
[240] Redepenning J, Venkataraman G, Chen J, Stafford N. Electrochemical preparation of chitosan/hydroxyapatite composite coatings on titanium substrates. *J. Biomed. Mater. Res. A* 2003;66A:411-416.
[241] Rhee SH, Tanaka J. Synthesis of a hydroxyapatite/collagen/chondroitin sulfate nanocomposite by a novel precipitation method. *J. Am. Ceram. Soc.* 2001;84:459-461.
[242] Pezzotti G, Asmus SMF. Fracture behavior of hydroxyapatite/polymer interpenetrating network composites prepared by *in situ* polymerization process. *Mater. Sci. Eng. A* 2001;316:231-237.
[243] Kealley C, Ben-Nissan B, van Riessen A, Elcombe M. Development of carbon nanotube reinforced hydroxyapatite bioceramics. *Key Engin. Mater.* 2006;309-311:597-600.
[244] Kealley C, Elcombe M, van Riessen A, Ben-Nissan B. Development of carbon nanotube reinforced hydroxyapatite bioceramics. *Physica B* 2006;385-386:496-498.

[245] Aryal S, Bahadur KCR, Dharmaraj N, Kim KW, Kim HY. *Synthesis and characterization of hydroxyapatite using carbon nanotubes as a nanomatrix. Scripta Mater.* 2006;54:131-135.

[246] Wei Q, Yang XP, Chen GQ, Tang JT, Deng XL. The ultrasonic assisted synthesis of nano-hydroxyapatite and MWNT/hydroxyapatite composites. *New Carbon Mater.* 2005;20:164-170.

[247] Zhao LP, Gao L. Novel *in situ* synthesis of MWNT-hydroxyapatite composites. *Carbon* 2004;42:423-460.

[248] Aryal S, Bhattarai SR, Bahadur KCR, Khil MS, Lee DR, Kim HY. Carbon nanotubes assisted biomimetic synthesis of hydroxyapatite from simulated body fluid. *Mater. Sci. Eng. A* 2006;426:202-207.

[249] Rautaray D, Mandal S, Sastry M. Synthesis of hydroxyapatite crystals using amino acid-capped gold nanoparticles as a scaffold. *Langmuir* 2005;21:5185-5191.

[250] Memoto R, Nakamura S, Isobe T, Senna M. Direct synthesis of hydroxyapatite-silk fibroin nano-composite sol via a mechano-chemical route. *J. Sol Gel Sci. Technol.* 2001;21:7-12.

[251] Fujiwara M, Shiokawa K, Morigaki K, Tatsu Y, Nakahara Y. Calcium phosphate composite materials including inorganic powders, BSA or duplex DNA prepared by W/O/W interfacial reaction method. *Mater. Sci. Eng. C* 2008;28:280-288.

[252] Nagata F, Miyajima T, Yokogawa Y. A method to fabricate hydroxyapatite/poly(lactic acid) microspheres intended for biomedical application. *J. Eur. Ceram. Soc.* 2006;26:533-535.

[253] Russias J, Saiz E, Nalla RK, Tomsia AP. Microspheres as building blocks for hydroxyapatite/polylactide biodegradable composites. *J. Mater. Sci.* 2006;41:5127-5133.

[254] Khan YM, Cushnie EK, Kelleher JK, Laurencin CT. *In situ* synthesized ceramic-polymer composites for bone tissue engineering: bioactivity and degradation studies. *J. Mater. Sci.* 2007;42:4183-4190.

[255] Kim HW, Knowles JC, Kim HE. Hydroxyapatite and gelatin composite foams processed via novel freeze-drying and crosslinking for use as temporary hard tissue scaffolds. *J. Biomed. Mater. Res. A* 2005;72A:136-145.

[256] Sinha A, Das G, Sharma BK, Roy RP, Pramanick AK, Nayar S. Poly(vinyl alcohol)-hydroxyapatite biomimetic scaffold for tissue regeneration. *Mater. Sci. Eng. C* 2007;27:70-74.

[257] Sun L, Berndt CC, Gross KA. Hydroxyapatite/polymer composite flame-sprayed coatings for orthopedic applications. *J. Biomater. Sci. Polym. Edn.* 2002;13:977-990.

[258] Sugawara A, Yamane S, Akiyoshi K. Nanogel-templated mineralization: polymer-calcium phosphate hybrid nanomaterials. *Macromol. Rapid Commun.* 2006;27:441-446.

[259] Liu Q, de Wijn JR, van Blitterswijk CA. Nanoapatite/polymer composites: mechanical and physicochemical characteristics. *Biomaterials* 1997;18:1263-1270.

[260] Uskokovic PS, Tang CY, Tsui CP, Ignjatovic N, Uskokovic DP. Micromechanical properties of a hydroxyapatite/poly-L-lactide biocomposite using nanoindentation and modulus mapping. *J. Eur. Ceram. Soc.* 2007;27:1559-1564.

[261] Todo M, Kagawa T. Improvement of fracture energy of HA/PLLA biocomposite material due to press processing. *J. Mater. Sci.* 2008;43:799-801.

[262] Woo KM, Seo J, Zhang RY, Ma PX. Suppression of apoptosis by enhanced protein adsorption on polymer/hydroxyapatite composite scaffolds. *Biomaterials* 2007;28:2622-2630.

[263] Ma PX, Zhang R, Xiao G, Franceschi R. Engineering new bone tissue *in vitro* on highly porous poly(α-hydroxyl acids)/hydroxyapatite composite scaffolds. *J. Biomed. Mater. Res.* 2001;54:284-293.

[264] Wang M, Chen LJ, Ni J, Weng J, Yue CY. Manufacture and evaluation of bioactive and biodegradable materials and scaffolds for tissue engineering. *J. Mater. Sci. Mater. Med.* 2001;12:855-860.

[265] Baji A, Wong SC, Srivatsan TS, Njus GO, Mathur G. Processing methodologies for polycaprolactone-hydroxyapatite composites: a review. Mater. *Manuf. Process.* 2006;21:211-218.

[266] Wei G, Ma PX. Macroporous and nanofibrous polymer scaffolds and polymer/bone-like apatite composite scaffolds generated by sugar spheres. *J. Biomed. Mater. Res.* A 2006;78A:306-315.

[267] Guan L, Davies JE. Preparation and characterization of a highly macroporous biodegradable composite tissue engineering scaffold. *J. Biomed. Mater. Res. A* 2004;71A:480-487.

[268] Teng XR, Ren J, Gu SY. Preparation and characterization of porous PDLLA/HA composite foams by supercritical carbon dioxide technology. *J. Biomed. Mater. Res. B Appl. Biomater.* 2007;81B:185-193.

[269] Ren J, Zhao P, Ren T, Gu S, Pan K. Poly (D,L-lactide)/nano-hydroxyapatite composite scaffolds for bone tissue engineering and biocompatibility evaluation. *J. Mater. Sci. Mater. Med.* 2008;19:1075-1082.
[270] Wang M, Yue CY, Chua B. Production and evaluation of hydroxyapatite reinforced polysulfone for tissue replacement. *J. Mater. Sci. Mater. Med.* 2001;12:821-826.
[271] Chlopek J, Rosol P, Morawska-Chochol A. Durability of polymer-ceramics composite implants determined in creep tests. *Compos. Sci. Technol.* 2006;66:1615-1622.
[272] Szaraniec B, Rosol P, Chlopek J. Carbon composite material and polysulfone modified by nano-hydroxyapatite. *e-Polymers* 2005;030:1-7.
[273] Nayar S, Sinha A. Systematic evolution of a porous hydroxyapatite-poly(vinylalcohol)-gelatin composite. *Colloids Surf. B Biointerfaces* 2004;35:29-32.
[274] Chang MC, Ko CC, Douglas WH. Modification of hydroxyapatite/gelatin composite by polyvinylalcohol. *J. Mater. Sci.* 2005;40:505-509.
[275] Chang MC, Ko CC, Douglas WH. Modification of hydroxyapatite/gelatin composite by polyvinylalcohol. *J. Mater. Sci.* 2005;40:2723-2727.
[276] You C, Miyazaki T, Ishida E, Ashizuka M, Ohtsuki C, Tanihara M. Fabrication of poly(vinyl alcohol)-apatite hybrids through biomimetic process. *J. Eur. Ceram. Soc.* 2007;27:1585-1588.
[277] Xu F, Li Y, Yao X, Liao H, Zhang L. Preparation and *in vivo* investigation of artificial cornea made of nano-hydroxyapatite/poly (vinyl alcohol) hydrogel composite. *J. Mater. Sci. Mater. Med.* 2007;18:635-640.
[278] Xu F, Li Y, Deng Y, Xiong G. Porous nano-hydroxyapatite/poly(vinyl alcohol) composite hydrogel as artificial cornea fringe: characterization and evaluation *in vitro*. *J. Biomater. Sci. Polymer Edn.* 2008;19:431-439.
[279] Nayar S, Pramanick AK, Sharma BK, Das G, Kumar BR, Sinha A. Biomimetically synthesized polymer-hydroxyapatite sheet like nano-composite. *J. Mater. Sci. Mater. Med.* 2008;19:301-304.
[280] Bigi A, Boanini E, Gazzano M, Rubini K. Structural and morphological modifications of hydroxyapatite-polyaspartate composite crystals induced by heat treatment. *Cryst. Res. Technol.* 2005;40:1094-1098.

[281] Bertoni E, Bigi A, Falini G, Panzavolta S, Roveri N. Hydroxyapatite polyacrylic acid nanocrystals. *J. Mater. Chem.* 1999;9:779-782.
[282] Pramanik N, Biswas SK, Pramanik P. Synthesis and characterization of hydroxyapatite/poly(vinyl alcohol phosphate) nanocomposite biomaterials. *Int. J. Appl. Ceram. Technol.* 2008;5:20-28.
[283] Qiu HJ, Yang J, Kodali P, Koh J, Ameer GA. A citric acid-based hydroxyapatite composite for orthopedic implants. *Biomaterials* 2006;27:5845-5854.
[284] Greish YE, Brown PW. Chemically formed HAp-Ca poly(vinyl phosphonate) composites. *Biomaterials* 2001;22:807-816.
[285] Greish YE, Brown PW. Preparation and characterization of calcium phosphate-poly(vinyl phosphonic acid) composites. *J. Mater. Sci. Mater. Med.* 2001;12:407-411.
[286] Greish YE, Brown PW. Formation and properties of hydroxyapatite-calcium poly(vinyl phosphonate) composites. *J. Am. Ceram. Soc.* 2002;85:1738-1744.
[287] Nakahira A, Tamai M, Miki S, Pezotti G. Fracture behavior and biocompatibility evaluation of nylon-infiltrated porous hydroxyapatite. *J. Mater. Sci.* 2002;37:4425-4430.
[288] Sailaja GS, Velayudhan S, Sunny MC, Sreenivasan K, Varma HK, Ramesh P. Hydroxyapatite filled chitosan-polyacrylic acid polyelectrolyte complexes. *J. Mater. Sci.* 2003;38:3653-3662.
[289] Zhang H, Xu JJ, Chen HY. Electrochemically deposited 2D nanowalls of calcium phosphate-PDDA on a glassy carbon electrode and their applications in biosensing. *J. Phys. Chem. C* 2007;111:16564-16570.
[290] Piticescu RM, Chitanu GC, Albulescu M, Giurginca M, Popescu ML, Łojkowski W. Hybrid HAp-maleic anhydride copolymer nanocomposites obtained by *in-situ* functionalisation. *Solid State Phenomena* 2005;106:47-56.
[291] Enlow D, Rawal A, Kanapathipillai M, Schmidt-Rohr K, Mallapragada S, Lo CT, Thiyagarajan P, Akin M. Synthesis and characterization of self-assembled block copolymer templated calcium phosphate nanocomposite gels. *J. Mater. Chem.* 2007;17:1570-1578.
[292] Kaito T, Myoui A, Takaoka K, Saito N, Nishikawa M, Tamai N, Ohgushi H, Yoshikawa H. Potentiation of the activity of bone morphogenetic protein-2 in bone regeneration by a PLA-PEG/hydroxyapatite composite. *Biomaterials* 2005;26:73-79.
[293] Song J, Saiz E, Bertozzi CR. A new approach to mineralization of biocompatible hydrogel scaffolds: an efficient process toward 3-

dimensional bonelike composites. *J. Am. Chem. Soc.* 2003;125:1236-1243.
[294] Meenan BJ, McClorey C, Akay M. Thermal analysis studies of poly(etheretherketone)/hydroxyapatite biocomposite mixtures. *J. Mater. Sci. Mater. Med.* 2000;11:481-489.
[295] Fan JP, Tsui CP, Tang CY, Chow CL. Modeling of the mechanical behavior of HA/PEEK biocomposite under quasi-static tensile load. *Biomaterials* 2004;25:5363-5373.
[296] Abu Bakar MS, Cheng MHW, Tang SM, Yu SC, Liao K, Tan CT, Khor KA, Cheang P. Tensile properties, tension-tension fatigue and biological response of polyetheretherketone-hydroxyapatite composites for load-bearing orthopedic implants. *Biomaterials* 2003;24:2245-2250.
[297] Abu Bakar MS, Cheang P, Khor KA. Mechanical properties of injection molded hydroxyapatite-polyetheretherketone biocomposites. *Compos. Sci. Technol.* 2003;63:421-425.
[298] Abu Bakar MS, Cheang P, Khor KA. Tensile properties and microstructural analysis of spheroidized hydroxyapatite-poly(etheretherketone) biocomposites. *Mater. Sci. Eng. A* 2003;345:55-63.
[299] Fan JP, Tsui CP, Tang CY. Modeling of the mechanical behavior of HA/PEEK biocomposite under quasi-static tensile load. *Mater. Sci. Eng. A* 2004;382:341-350.
[300] Yu S, Hariram KP, Kumar R, Cheang P, Aik KK. *In vitro* apatite formation and its growth kinetics on hydroxyapatite/polyetheretherketonebiocomposites. *Biomaterials* 2005;26:2343-2352.
[301] Gong XH, Tang CY, Hu HC, Zhou XP. Improved mechanical properties of HIPS/hydroxyapatite composites by surface modification of hydroxyapatite via *in situ* polymerization of styrene. *J. Mater. Sci. Mater. Med.* 2004;15:1141-1146.
[302] Laurencin CT, Attawia MA, Elgendy HE, Herbert KM. Tissue engineered bone-regeneration using degradable polymers: the formation of mineralized matrices. *Bone* 1996;91:S93-S99.
[303] Laurencin CT, Attawia MA, Lu LQ, Borden MD, Lu HH, Gorum WJ, Lieberman JR. Poly(lactide-*co*-glycolide)/hydroxyapatite delivery of BMP-2-producing cells: a regional gene therapy approach to bone regeneration. *Biomaterials* 2001;22:1271-1277.

[304] Kim SS, Ahn KM, Park MS, Lee JH, Choi CY, Kim BS. A poly(lactide-co-glycolide)/hydroxyapatite composite scaffold with enhanced osteoconductivity. *J. Biomed. Mater. Res. A* 2007;80A:206-215.

[305] Oliveira J, Miyazaki T, Lopes M, Ohtsuki C, Santos J. Bonelike®/PLGA hybrid materials for bone regeneration: preparation route and physicochemical characterization. *J. Mater. Sci. Mater. Med.* 2005;16:253-259.

[306] Kim S, Kim SS, Lee SH, Ahn SE, Gwak SJ, Song JH, Kim BS, Chung HM. In vivo bone formation from human embryonic stem cell-derived osteogenic cells in poly(D,L-lactic-co-glycolic acid)/hydroxyapatite composite scaffolds. *Biomaterials* 2008;29:1043-1053.

[307] Petricca SE, Marra KG, Kumta PN. Chemical synthesis of poly(lactic-co-glycolic acid)/hydroxyapatite composites for orthopaedic applications. *Acta Biomater.* 2006;2:277-286.

[308] Sato M, Slamovich EB, Webster TJ. Enhanced osteoblast adhesion on hydrothermally treated hydroxyapatite/titania/poly(lactide-co-glycolide) sol-gel titanium coatings. *Biomaterials* 2005;26:1349-1357.

[309] Gu SY, Zhan H, Ren J, Zhou XY. Sol-gel synthesis and characterisation of nano-sized hydroxyapatite powders and hydroxyapatite/poly(D,L-lactide-co-glycolide) composite scaffolds. *Polymers & Polymer Composites* 2007;15:137-144.

[310] Verheyen CCPM, de Wijin JR, van Blitterwijk CA, de Groot K. Evaluation of hydroxyapatite/poly(L-lactide) composites: mechanical behavior. *J. Biomed. Mater. Res.* 1992;26:1277-1296.

[311] Balac I, Uskokovic PS, Aleksic R, Uskokovic D. Predictive modeling of the mechanical properties of particulate hydroxyapatite reinforced polymer composites. *J. Biomed. Mater. Res.* 2002;63:793-799.

[312] Dawes E, Rushton N. The effects of lactic acid on PGE2 production by macrophages and human synovial fibroblasts: a possible explanation for problems associated with the degradation of poly(lactide) implants? *Clin. Mater.* 1994;17:157-163.

[313] Verheyen CCPM, Klein CPAT, de Blieck-Hogervorst JMA, Wolke JGC, de Wijn JR, van Blitterswijk CA, de Groot K. Evaluation of hydroxylapatite poly(L-lactide) composites-physicochemical properties. *J. Mater. Sci. Mater. Med.* 1993;4:58-65.

[314] Li H, Chang J. pH-compensation effect of bioactive inorganic fillers on the degradation of PLGA. *Compos. Sci. Technol.* 2005;65:2226-2232.

[315] Agrawal CM, Athanasiou KA. Technique to control pH in vicinity of biodegrading PLA-PGA implants. *J. Biomed. Mater. Res. Appl. Mater.* 1997;38:105-114.
[316] Peter SJ, Miller ST, Zhu G, Yasko AW, Mikos AG. *In vivo* degradation of a poly(propylene fumarate)/β-tricalcium phosphate injectable composite scaffold. *J. Biomed. Mater. Res.* 1998;41:1-7.
[317] Ara M, Watanabe M, Imai Y. Effect of blending calcium compounds on hydrolitic degradation of poly(D,L-lactic acid-*co*-glycolic acid). *Biomaterials* 2002;23:2479-2483.
[318] Linhart W, Peters F, Lehmann W, Schwarz K. Schilling A, Amling M. Rueger JM, Epple M. Biologically and chemically optimized composites of carbonated apatite and polyglycolide as bone substitution materials. *J. Biomed. Mater. Res.* 2001;54:162-171.
[319] Schiller C, Epple M. Carbonated apatites can be used as pH-stabilizing filler for biodegradable polyesters. *Biomaterials* 2003;24:2037-2043.
[320] Schiller C, Rasche C, Wehmöller M, Beckmann F, Eufinger H, Epple M, Weihe S. Geometrically structured implants for cranial reconstruction made of biodegradable polyesters and calcium phosphate/calcium carbonate. *Biomaterials* 2004;25:1239-1247.
[321] Shikinami Y, Okuno M. Bioresorbable devices made of forged composites of hydroxyapatite (HA) particles and poly L-lactide (PLLA). Part II: practical properties of miniscrews and miniplates. *Biomaterials* 2001;22:3197-3211.
[322] Russias J, Saiz E, Nalla RK, Gryn K, Ritchie RO, Tomsia AP. Fabrication and mechanical properties of PLA/HA composites: a study of *in vitro* degradation. *Mater. Sci. Eng. C* 2006;26:1289-1295.
[323] Kim HW, Lee HH, Knowles JC. Electrospinning biomedical nanocomposite fibers of hydroxyapaite/poly(lactic acid) for bone regeneration. *J. Biomed. Mater. Res. A* 2006;79A:643-649.
[324] Gross KA, Rodríguez-Lorenzo LM. Biodegradable composite scaffolds with an interconnected spherical network for bone tissue engineering. *Biomaterials* 2004;25:4955-4962.
[325] Durucan C, Brown PW. Low temperature formation of calcium-deficient hydroxyapatite-PLA/PLGA composites. *J. Biomed. Mater. Res.* 2000;51:717-725.
[326] Durucan C, Brown PW. Calcium-deficient hydroxyapatite-PLGA composites: mechanical and microstructural investigation. *J. Biomed. Mater. Res.* 2000;51:726-734.

[327] Ignjatovic N, Suljovrujic E, Biudinski-Simendic J, Krakovsky I, Uskokovic D. Evaluation of hot-presses hydroxyapatite/poly-L-lactide composite biomaterial characteristics. *J. Biomed. Mater. Res. B Appl. Biomater.* 2004;71B:284-294.

[328] Nazhat SN, Kellomäki M, Törmälä P, Tanner KE, Bonfield W. Dynamic mechanical characterization of biodegradable composites of hydroxyapatite and polylactides. *J. Biomed. Mater. Res.* 2001;58:335-343.

[329] Hasegawa S, Tamura J, Neo M, Goto K, Shikinami Y, Saito M, Kita M, Nakamura T. In vivo evaluation of a porous hydroxyapatite/poly-DL-lactide composite for use as a bone substitute. *J. Biomed. Mater. Res. A* 2005;75A:567-579.

[330] Hasegawa S, Neo M, Tamura J, Fujibayashi S, Takemoto M, Shikinami Y, Okazaki K, Nakamura T. In vivo evaluation of a porous hydroxyapatite/poly-DL-lactide composite for bone tissue engineering. *J. Biomed. Mater. Res. A* 2007;81A:930-938.

[331] Higashi S, Yamamuro T, Nakamura T, Ikada Y, Hyon SH, Jamshidi K. Polymer-hydroxyapatite composites for biodegradable bone fillers. *Biomaterials* 1986;7:183-187.

[332] Ylinen P. Filling of bone defects with porous hydroxyapatite reinforced with polylactide or polyglycolide fibres. *J. Mater. Sci. Mater. Med.* 1994;5:522-528.

[333] Reis RL, Cunha AM, Bevis MJ. Using nonconventional processing routes to develop anisotropic and biodegradable composites of starch-based thermoplastics reinforced with bone-like ceramics. *J. Appl. Med. Polym.* 1998;2:49-53.

[334] Reis RL, Cunha AM. New degradable load-bearing biomaterials composed of reinforced starch based blends. *J. Appl. Med. Polym.* 2000;4:1-5.

[335] Sousa RA, Mano JF, Reis RL, Cunha AM, Bevis MJ. Mechanical performance of starch based bioactive composites moulded with preferred orientation for potential medical applications. *Polym. Eng. Sci.* 2002;42:1032-1045.

[336] Marques AP, Reis RL. Hydroxyapatite reinforcement of different starch-based polymers affects osteoblast-like cells adhesion/spreading and proliferation. *Mater. Sci. Eng. C* 2005;25:215-229.

[337] Reis RL, Cunha AM, Allan PS, Bevis MJ. Structure development and control of injection-molded hydroxylapatite-reinforced starch/EVOH composites. *J. Polym. Adv. Tech.* 1997;16:263-277.

[338] Vaz CM, Reis RL, Cunha AM. Use of coupling agents to enhance the interfacial interactions in starch-EVOH/hydroxylapatite composites. *Biomaterials* 2002;23:629-635.
[339] Leonor IB, Ito A, Onuma K, Kanzaki N, Reis RL. *In vitro* bioactivity of starch thermoplastic/hydroxyapatite composite biomaterials: an *in situ* study using atomic force microscopy. *Biomaterials* 2003;24:579-585.
[340] Vaz CM, Reis RL, Cunha AM. Degradation model of starch-EVOH+HA composites. *Mat. Res. Innovat.* 2001;4:375-380.
[341] Boeree N, Dove J, Cooper JJ, Knowles JC, Hastings GW. Development of a degradable composite for orthopaedic use – mechanical evaluation of an hydroxyapatite-polyhydroxybutyrate composite materials. *Biomaterials* 1993;14:793-796.
[342] Doyle C, Tanner ET, Bonfield W. *In vitro* and *in vivo* evaluation of polyhydroxybutyrate and of polyhydroxybutyrate reinforced with hydroxyapatite. *Biomaterials* 1991;12:841-847.
[343] Chen LJ, Wang M. Production and evaluation of biodegradable composites based on PHB-PHV copolymer. *Biomaterials* 2002;23:2631-2639.
[344] Ni J, Wang M. *In vitro* evaluation of hydroxyapatite reinforced polyhydroxybutyrate composite. *Mater. Sci. Eng. C* 2002;20:101-109.
[345] Knowles JC, Hastings GW, Ohta H, Niwa S, Boeree N. Development of a degradable composite for orthopedic use – *in vivo* biomechanical and histological evaluation of two bioactive degradable composites based on the polyhydroxybutyrate polymer. *Biomaterials* 1992;13:491-496.
[346] Luklinska ZB, Bonfield W. Morphology and ultrastructure of the interface between hydroxyapatite-polybutyrate composite implant and bone. *J. Mater. Sci. Mater. Med.* 1997;8:379-383.
[347] Chen DZ, Tang CY, Chan KC, Tsui CP, Yu PHF, Leung MCP, Uskokovic PS. Dynamic mechanical properties and *in vitro* bioactivity of PHBHV/HA nanocomposite. *Compos. Sci. Technol.* 2007;67:1617-1626.
[348] Rai B, Noohom W, Kithva PH, Grøndahl L, Trau M. Bionanohydroxyapatite/poly(3-hydroxybutyrate-co-3-hydroxyvalerate) composites with improved particle dispersion and superior mechanical properties. *Chem. Mater.* 2008;20:2802-2808.
[349] Wang YW, Wu Q, Chen J, Chen GQ. Evaluation of three-dimensional scaffolds made of blends of hydroxyapatite and poly(3-hydroxybutyrate-*co*-3-hydroxyhexanoate) for bone reconstruction. *Biomaterials* 2005;26:899-904.

[350] Linhart W, Lehmann W, Siedler M, Peters F, Schilling AF, Schwarz K, Amling M, Rueger JM, Epple M. Composites of amorphous calcium phosphate and poly(hydroxybutyrate) and poly(hydroxybutyrate-*co*-hydroxyvalerate) for bone substitution: assessment of the biocompatibility. *J. Mater. Sci.* 2006;41:4806-4813.

[351] Azevedo M, Reis RL, Claase M, Grijpma D, Feijen J. Development and properties of polycaprolactone/hydroxyapatite composite biomaterials. *J. Mater. Sci. Mater. Med.* 2003;14:103-107.

[352] Choi D, Marra KG, Kumta PN. Chemical synthesis of hydroxyapatite/poly(ε-caprolactone) composites. *Mater. Res. Bull.* 2004;39:417-432.

[353] Hao J, Yuan M, Deng X. Biodegradable and biocompatible nanocomposites of poly(ε-caprolactone) with hydroxyapatite nanocrystals: thermal and mechanical properties. *J. Appl. Polymer Sci.* 2003;86:676-683.

[354] Walsh D, Furuzono T, Tanaka J. Preparation of porous composite implant materials by *in situ* polymerization of porous apatite containing ε-caprolactone or methyl methacrylate. *Biomaterials* 2001;22:1205-1212.

[355] Cerrai P, Guerra GD, Tricoli M, Krajewski A, Guicciardi S, Ravaglioli A, Maltinti S, Masetti G. New composites of hydroxyapatite and bioresorbable macromolecular material. *J. Mater. Sci. Mater. Med.* 1999;10:283-289.

[356] Kim HW. Biomedical nanocomposites of hydroxyapatite/polycaprolactone obtained by surfactant mediation. *J. Biomed. Mater. Res. A* 2007;83A:169-177.

[357] Kim HW, Knowles JC, Kim HE. Development of hydroxyapatite bone scaffold for controlled drug release via poly(ε-caprolactone) and hydroxyapatite hybrid coatings. *J. Biomed. Mater. Res. A* 2004;70A:240-249.

[358] Guerra GD, Cerrai P, Tricoli M, Krajewski A, Ravaglioli A, Mazzocchi M, Barbani N. Composites between hydroxyapatite and poly(ε-caprolactone) synthesized in open system at room temperature. *J. Mater. Sci. Mater. Med.* 2006;17:69-79.

[359] Causa F, Netti PA, Ambrosio L, Ciapetti G, Baldini N, Pagani S, Martini D, Giunti A. Poly-ε-caprolactone/hydroxyapatite composites for bone regeneration: *in vitro* characterization and human osteoblast response. *J. Biomed. Mater. Res. A* 2006;76A:151-162.

[360] Thomas V, Jagani S, Johnson K, Jose MV, Dean DR, Vohra YK, Nyairo E. Electrospun bioactive nanocomposite scaffolds of polycaprolactone and nanohydroxyapatite for bone tissue engineering. *J. Nanosci. Nanotechol.* 2006;6:487-493.

[361] Dunn A, Campbell P, Marra KG. The influence of polymer blend composition on the degradation of polymer/hydroxyapatite biomaterials. *J. Mater. Sci. Mater. Med.* 2001;12:673-677.

[362] Calandrelli L, Immirzi B, Malinconico M, Volpe M, Oliva A, Ragione F. Preparation and characterization of composites based on biodegradable polymers for *in vivo* application. *Polymer* 2000;41:8027-8033.

[363] Chen B, Sun K. Poly(ε-caprolactone)/hydroxyapatite composites: effects of particle size, molecular weight distribution and irradiation on interfacial interaction and properties. *Polymer Testing* 2005;24:64-70.

[364] Ural E, Kesenci K, Fambri L, Migliaresi C, Piskin E. Poly(D,L-lactide/ε-caprolactone)/hydroxyapatite composites. *Biomaterials* 2000;21:2147-2154.

[365] Yoon BH, Kim HW, Lee SH, Bae CJ, Koh YH, Kong YM, Kim HE. Stability and cellular responses to fluorapatite-collagen composites. *Biomaterials* 2005;26:2957-2963.

[366] Kim HW, Lee EJ, Kim HE, Salih V, Knowles JC. Effect of fluoridation of hydroxyapatite in hydroxyapatite/polycaprolactone composites on osteoblast activity. *Biomaterials* 2005;26:4395-4404.

[367] Busch S, Dolhaine H, DuChesne A, Heinz S, Hochrein O, Laeri F, Podebrad O, Vietze U, Weiland T, Kniep R. Biomimetic morphogenesis of fluorapatite-gelatin composites: fractal growth, the question of intrinsic electric fields, core/shell assemblies, hollow spheres and reorganization of denatured collagen. *Eur. J. Inorg. Chem.* 1999;1643-1653.

[368] Busch S, Schwarz U, Kniep R. Chemical and structural investigations of biomimetically grown fluorapatite-gelatin composite aggregates. *Adv. Funct. Mater.* 2003;13:189-198.

[369] Simon P, Carrillo-Cabrera W, Formanek P, Göbel C, Geiger D, Ramlau R, Tlatlik H, Buder J, Kniep R. On the real-structure of biomimetically grown hexagonal pristamic seed of fluoroapatite-gelatin-composites: TEM investigations along [001]. *J. Mater. Chem.* 2004;14:2218-2224.

[370] Göbel C, Simon P, Buder J, Tlatlik H, Kniep R. Phase formation and morphology of calcium phosphate-gelatin-composites grown by double

diffusion technique: the influence of fluoride. *J. Mater. Chem.* 2004;14:2225-2230.
[371] Simon P, Schwarz U, Kniep R. Hierarchical architecture and real structure in a biomimetic nano-composite of fluorapatite with gelatine: a model system for steps in dentino- and osteogenesis? *J. Mater. Chem.* 2005;15:4992-4996.
[372] Tlatlik H, Simon P, Kawska A, Zahn D, Kniep R. Biomimetic fluorapatite-gelatin nanocomposites: pre-structuring of gelatin matrices by ion impregnation and its effect on form development. *Angew. Chem. Int. Ed. Engl.* 2006;45:1905-1910.
[373] Simon P, Zahn D, Lichte H, Kniep R. Intrinsic electric dipole fields and the induction of hierarchical form developments in fluorapatite-gelatin nanocomposites: a general principle for morphogenesis of biominerals? *Angew. Chem. Int. Ed. Engl.* 2006;45:1911-1915.
[374] Kniep R, Simon P. Fluorapatite-gelatin-nanocomposites: self-organized morphogenesis, real structure and relations to natural hard materials. In: Crystallization and self-organization process. (Eds.) Naka, Kensuke. *Topics in current chemistry 270.* Springer, Berlin 2007;73-125.
[375] Kniep R, Simon P. "Hidden" hierarchy of microfibrils within 3D-periodic fluorapatite-gelatin nanocomposites: development of complexity and form in a biomimetic system. *Angew. Chem. Int. Ed. Engl.* 2008;47:1405-1409.
[376] Handschel J, Wiesmann HP, Stratmann U, Kleinheinz J, Meyer U, Joos U. TCP is hardly resorbed and not osteoconductive in a non-loading calvarial model. *Biomaterials* 2002;23:1689-1695.
[377] Kikuchi M, Tanaka J, Koyama Y, Takakuda K. Cell culture tests of TCP/CPLA composites. *J. Biomed. Mater. Res.* 1999;48:108-110.
[378] Yaszemski MJ, Payne RG, Hayes WC, Langer R, Mikos AG. *In vitro* degradation of a poly(propylene fumarate)-based composite material. *Biomaterials* 1996;17:2117-2130.
[379] Wang M, Wang J, Ni J. Developing tricalcium phosphate/polyhydroxybutyrate composite as a new biodegradable material for clinical applications. *Biomechanics* 2000;192:741-744.
[380] Kikuchi M, Koyama Y, Takakuda K, Miyairi H, Shirahama N, Tanaka J. *In vitro* change n mechanical strength of β-tricalcium phosphate/copolymerized poly-L-lactide composites and their application for guided bone regeneration. *J. Biomed. Mater. Res.* 2002;62:265-272.

[381] Ignatius AA, Augat P, Claes LE. Degradation behaviour of composite pins made of tricalcium phosphate and poly(L,DL-lactide). *J. Biomater. Sci. Polym. Edn.* 2001;12:185-194.

[382] Ignatius AA, Wolf S, Augat P, Claes LE. Composites made of rapidly resorbable ceramics and poly(lactide) show adequate mechanical properties for use as bone substitute materials. *J. Biomed. Mater. Res.* 2001;57:126-131.

[383] Kikuchi M, Tanaka J. Chemical interaction in β-tricalcium phosphate/copolymerized poly-L-lactide composites. *J. Ceram. Soc. Japan.* 2000;108:642-645.

[384] Aunoble S, Clement D, Frayssinet P, Harmand MF, le Huec JC. Biological performance of a new β-TCP/PLLA composite material for applications in spine surgery: *in vitro* and *in vivo* studies. *J. Biomed. Mater. Res. A* 2006;78A:416-422.

[385] Kikuchi M, Koyama Y, Yamada T, Imamura Y, Okada T, Shirahama N, Akita K, Takakuda K, Tanaka J. Development of guided bone regeneration membrane composed of β-tricalcium phosphate and poly(L-lactide-*co*-glycolide-*co*-ε-caprolactone) composites. *Biomaterials* 2004;25:5979-5986.

[386] Chen TM, Yao CH, Wang HJ, Chou GH, Lee TW, Lin FH. Evaluation of a novel malleable, biodegradable osteoconductive composite in a rabbit cranial defect model. *Mater. Chem. Phys.* 1998;55:44-50.

[387] Dong GC, Chen HM, Yao CH. A novel bone substitute composite composed of tricalcium phosphate, gelatin and drynaria fortunei herbal extract. *J. Biomed. Mater. Res. A* 2008;84A:167-177.

[388] Yao CH, Liu BS, Hsu SH, Chen YS, Tsai CC. Biocompatibility and biodegradation of a bone composite containing tricalcium phosphate and genipin crosslinked gelatin. *J. Biomed. Mater. Res. A* 2004;69A:709-717.

[389] Lin FH, Yao CH, Sun JS, Liu HC, Huang CW. Biological effects and cytotoxicity of the composite composed by tricalcium phosphate and glutaraldehyde cross-linked gelatin. *Biomaterials* 1998;19:905-917.

[390] Eslaminejad MB, Mirzadeh H, Mohamadi Y, Nickmahzar A. Bone differentiation of marrow-derived mesenchymal stem cells using β-tricalcium phosphate-alginate-gelatin hybrid scaffolds. *J. Tissue Eng. Regen. Med.* 2007;1:417-424.

[391] Takahashi Y, Yamamoto M, Tabata Y. Osteogenic differentiation of mesenchymal stem cells in biodegradable sponges composed of gelatin and β-tricalcium phosphate. *Biomaterials* 2005;26:3587-3596.

[392] Bigi A, Cantelli I, Panzavolta S, Rubini K. α-tricalcium phosphate-gelatin composite cements. *J. Appl. Biomat. Biomech.* 2004;2:81-87.
[393] Yang SH, Hsu CK, Wang KC, Hou SM, Lin FH. Tricalcium phosphate and glutaraldehyde crosslinked gelatin incorporating bone morphogenetic protein – a viable scaffold for bone tissue engineering. *J. Biomed. Mater. Res. B Appl. Biomater.* 2005;74B:468-475.
[394] Kato M, Namikawa T, Terai H, Hoshino M, Miyamoto S, Takaoka K. Ectopic bone formation in mice associated with a lactic acid/dioxanone/ethylene glycol copolymer-tricalcium phosphate composite with added recombinant human bone morphogenetic protein-2. *Biomaterials* 2006;27:3927-3933.
[395] Muramatsu K, Oba K, Mukai D, Hasegawa K, Masuda S, Yoshihara Y. Subacute systemic toxicity assessment of β-tricalcium phosphate/carboxymethyl-chitin composite implanted in rat femur. *J. Mater. Sci. Mater. Med.* 2007;18:513-522.
[396] Bleach NC, Tanner KE, Kellomäki M, Törmälä P. Effect of filler type on the mechanical properties of self-reinforced polylactide-calcium phosphate composites. *J. Mater. Sci. Mater. Med.* 2001;12:911-915.
[397] Liu L, Xiong Z, Yan YN, Hu YY, Zhang RJ, Wang SG. Porous morphology, porosity, mechanical properties of poly(α-hydroxy acid)-tricalcium phosphate composite scaffolds fabricated by low-temperature deposition. *J. Biomed. Mater. Res. A* 2007;82A:618-629.
[398] Zhang Y, Zhang MQ. Synthesis and characterization of macroporous chitosan/calcium phosphate composite scaffolds for tissue engineering. *J. Biomed. Mater. Res.* 2001;55:304-312.
[399] Rai B, Teoh SH, Hutmacher DW, Cao T, Ho KH. Novel PCL-based honeycomb scaffolds as drug delivery systems for rhBMP-2. *Biomaterials* 2005;26:3739-3748.
[400] Rai B, Teoh SH, Ho KH, Hutmacher DW, Cao T, Chen F, Yacob K. The effect of rhBMP-2 on canine osteoblasts seeded onto 3D bioactive polycaprolactone scaffolds. *Biomaterials* 2004;25:5499-5506.
[401] Lei Y, Rai B, Ho KH, Teoh SH. *In vitro* degradation of novel bioactive polycaprolactone-20% tricalcium phosphate composite scaffolds for bone engineering. *Mater. Sci. Eng. C* 2007;27:293-298.
[402] Miyai T, Ito A, Tamazawa G, Matsuno T, Sogo Y, Nakamura C, Yamazaki A, Satoh T. Antibiotic-loaded poly-ε-caprolactone and porous β-tricalcium phosphate composite for treating osteomyelitis. *Biomaterials* 2008;29:350-358.

[403] Takahashi Y, Yamamoto M, Tabata Y. Enhanced osteoinduction by controlled release of bone morphogenetic protein-2 from biodegradable sponge composed of gelatin and β-tricalcium phosphate. *Biomaterials* 2005;26:4856-4865.

[404] Ignatius AA, Betz O, Augat P, Claes LE. *In vivo* investigations on composites made of resorbable ceramics and poly(lactide) used as bone graft substitutes. *J. Biomed. Mater. Res. Appl. Biomater.* 2001;58:701-709.

[405] Miao X, Lim WK, Huang X, Chen Y. Preparation and characterization of interpenetrating phased TCP/HA/PLGA composites. *Mater. Lett.* 2005;59:4000-4005.

[406] Brodie JC, Goldie E, Connel G, Merry J, Grant MH. Osteoblast interactions with calcium phosphate ceramics modified by coating with type I collagen. *J. Biomed. Mater. Res. A* 2005;73A:409-421.

[407] Zhang LF, Sun R, Xu L, Du J, Xiong ZC, Chen HC, Xiong CD. Hydrophilic poly (ethylene glycol) coating on PDLLA/BCP bone scaffold for drug delivery and cell culture. *Mater. Sci. Eng. C* 2008;28:141-149.

[408] Ignjatovic N, Ninkov P, Ajdukovic Z, Konstantinovic V, Uskokovic D. Biphasic calcium phosphate/poly-(DL-lactide-*co*-glycolide) biocomposite as filler and blocks for reparation of bone tissue. *Mater. Sci. Forum* 2005;494:519-524.

[409] Ignjatovic N, Ninkov P, Ajdukovic Z, Vasiljevic-Radovic D, Uskokovic D. Biphasic calcium phosphate coated with poly-D,L-lactide-*co*-glycolide biomaterial as a bone substitute. *J. Eur. Ceram. Soc.* 2007;27:1589-1594.

[410] Ignjatovic N, Ninkov P, Kojic V, Bokurov M, Srdic V, Krnojelac D, Selakovic S, Uskokovic D. Cytotoxicity and fibroblast properties during *in vitro* test of biphasic calcium phosphate/poly-DL-lactide-*co*-glycolide biocomposites and different phosphate materials. *Microsc. Res. Techniq.* 2006;69:976-982.

[411] Ajdukovic Z, Ignjatovic N, Petrovic D, Uskokovic D. Substitution of osteoporotic alveolar bone by biphasic calcium phosphate/poly-DL-lactide-*co*-glycolide biomaterials. *J. Biomater. Appl.* 2007;21:317-328.

[412] Kim HW, Knowles JC, Kim HE. Effect of biphasic calcium phosphates on drug release and biological and mechanical properties of poly(ε-caprolactone) composite membranes. *J. Biomed. Mater. Res. A* 2004;70A:467-479.

[413] Matsuda A, Ikoma T, Kobayashi H, Tanaka J. Preparation and mechanical property of core-shell type chitosan/calcium phosphate composite fiber. *Mater. Sci. Eng. C* 2004;24:723-728.

[414] Tortet L, Gavarri JR, Nihoul G, Dianoux AJ. Proton mobilities in brushite and brushite/polymer composites. *Solid State Ionics* 1997;97:253-256.

[415] Tortet L, Gavarri JR, Musso J, Nihoul G, Sarychev AK. Percolation and modeling of proton conduction in polymer/brushite composites. *J. Solid State Chem.* 1998;141:392-403.

[416] Park MS, Eanes ED, Antonucci JM, Skrtic D. Mechanical properties of bioactive amorphous calcium phosphate/methacrylate composites. *Dent. Mater.* 1998;14:137-141.

[417] Skrtic D, Antonucci JM, Eanes ED, Eichmiller FC, Schumacher GE. Physicochemical evaluation of bioactive polymeric composites based on hybrid amorphous calcium phosphates. *J. Biomed. Mater. Res.* 2000;53:381-391.

[418] Skrtic D, Antonucci JM, Eanes ED, Eidelman N. Dental composites based on hybrid and surface-modified amorphous calcium phosphates. *Biomaterials* 2004;25:1141-1150.

[419] Skrtic D, Antonucci JM. Effect of bifunctional comonomers on mechanical strength and water sorption of amorphous calcium phosphate- and silanized glass-filled Bis-GMA-based composites. *Biomaterials* 2003;24:2881-2888.

[420] Gutierrez MC, Jobbágy M, Ferrer ML, del Monte F. Enzymatic synthesis of amorphous calcium phosphate-chitosan nanocomposites and their processing into hierarchical structures. *Chem. Mater.* 2008;20:11-13.

[421] Hakimimehr D, Liu DM, Troczynski T. *In-situ* preparation of poly(propylene fumarate) – hydroxyapatite composite. *Biomaterials* 2005;26:7297-7303.

[422] Perkin KK, Turner JL, Wooley KL, Mann S. Fabrication of hybrid nanocapsules by calcium phosphate mineralization of shell cross-linked polymer micelles and nanocages. *Nano Lett.* 2005;5:1457-1461.

[423] LeGeros RZ, Chohayeb A, Shulman A. Apatitic calcium orthophosphates: possible dental restorative materials. *J. Dent. Res.* 1982;61(Special issue):343.

[424] Brown WE, Chow LC. A new calcium orthophosphate setting cement. *J. Dent. Res.* 1983;62(Special issue):672.

[425] Brown WE, Chow LC. *Dental restorative cement pastes.* US Patent No. 4518430. May 21, 1985.
[426] Tas AC. Porous, biphasic $CaCO_3$-calcium phosphate biomedical cement scaffolds from calcite ($CaCO_3$) powder. *Int. J. Appl. Ceram. Technol.* 2007;4:152-163.
[427] Driskell TD, Heller AL, Koenigs JF. *Dental treatments.* US Patent No. 3913229. October 21, 1975.
[428] Dorozhkin SV. Calcium orthophosphate cements and concretes. *Materials* 2009;2:221-291.
[429] Barrelet JA, Hofmann M, Grover LM, Gbureck U. High-strength apatitic cement by modification with α-hydroxyacid salts. *Adv. Mater.* 2003;15:2091-2094.
[430] Bigi A, Bracci B, Panzavolta S. Effect of added gelatin on the properties of calcium phosphate cement. *Biomaterials* 2004;25:2893-2899.
[431] Bigi A, Panzavolta S, Sturba L, Torricelli P, Fini M, Giardino R. Normal and osteopenic bone-derived osteoblast response to a biomimetic gelatin-calcium phosphate bone cement. *J. Biomed. Mater. Res. A* 2006;78A:739-745.
[432] Panzavolta S, Torricelli P, Sturba L, Bracci B, Giardino R, Bigi A. Setting properties and *in vitro* bioactivity of strontium-enriched gelatin-calcium phosphate bone cements. *J. Biomed. Mater. Res. A* 2008;84A:965-972.
[433] Rammelt S, Neumann M, Hanisch U, Reinstorf A, Pompe W, Zwipp H, Biewener A. Osteocalcin enhances bone remodeling around hydroxyapatite/collagen composites. *J. Biomed. Mater. Res. A* 2005;73A:284-294.
[434] Wang X, Ye J, Wang Y, Chen L. Reinforcement of calcium phosphate cement by bio-mineralized carbon nanotube. *J. Am. Ceram. Soc.* 2007;90:962-964.
[435] *http://en.wikipedia.org/wiki/Concrete* (accessed in May 2009).
[436] Kim SB, Kim YJ, Yoon TL, Park SA, Cho IH, Kim EJ, Kim IA, Shin JW. The characteristics of a hydroxyapatite-chitosan-PMMA bone cement. *Biomaterials* 2004;25:5715-5723.
[437] Vallo CI, Montemartini PE, Fanovich MA, Lópes JMP, Cuadrado TR. Polymethylmethacrylate-based bone cement modified with hydroxyapatite. *J. Biomed. Mater. Res. Appl. Biomater.* 1999;48:150-158.

[438] Sogal A, Hulbert SF. Mechanical properties of a composite bone cement: polymethylmethacrylate and hydroxyapatite. *Bioceramics* 5. 1992, pp. 213-224.

[439] Harper EJ, Behiri JC, Bonfield W. Flexural and fatigue properties of a bone cement based upon polymethylmethacrylate and hydroxyapatite. *J. Mater. Sci. Mater. Med.* 1995;6:799-803.

[440] Harper EJ, Braden M, Bonfield W. Mechanical properties of hydroxyapatite reinforced poly(methylmethacrylate) bone cement after immersion in a physiological solution: influence of a silane coupling agent. *J. Mater. Sci. Mater. Med.* 2001;11:491-497.

[441] Moursi AM, Winnard AV, Winnard PL, Lannutti JJ, Seghi RR. Enhanced osteoblast response to a PMMA – HA composite. *Biomaterials* 2002;23:133-144.

[442] Dalby MJ, Di Silvio L, Harper EJ, Bonfield W. Initial interaction of osteoblasts with the surface of a hydroxyapatite-poly(mefhylmethacrylate) cement. *Biomaterials* 2001;22:1739-1747.

[443] Dalby MJ, Di Silvio L, Harper EJ, Bonfield W. Increasing hydroxyapatite incorporation into poly(methylmethacrylate) cement increases osteoblast adhesion and response. *Biomaterials* 2002;23:569-576.

[444] Itokawa H, Hiraide T, Moriya M, Fujimoto M, Nagashima G, Suzuki R, Fujimoto T. A 12 month *in vivo* study on the response of bone to a hydroxyapatite-polymethylmethacrylate cranioplasty composite. *Biomaterials* 2007;28:4922-4927.

[445] Cheang P, Khor KA. Effect of particulate morphology on the tensile behaviour of polymer-hydroxyapatite composites. *Mater. Sci. Eng. A* 2003;345:47-54.

[446] Dalby MJ, Di Silvio L, Harper EJ, Bonfield W. *In vitro* evaluation of a new polymethylmethacrylate cement reinforced with hydroxyapatite. *J. Mater. Sci. Mater. Med.* 1999;10:793-796.

[447] Deb S, Braden M, Bonfield W. Water absorption characteristics of modified hydroxyapatite bone cements. *Biomaterials* 1995;16:1095-1100.

[448] Borzacchiello A, Ambrosio L, Nicolais L. Harper EJ, Tanner KE, Bonfield W. Comparison between the polymerization behavior of a new bone cement and a commercial one: modeling and *in vitro* analysis. *J. Mater. Sci. Mater. Med.* 1998;9:835-838.

[449] Ohgaki M, Yamashita K. Preparation of polymethylmethacrylate-reinforced functionally graded hydroxyapatite composites. *J. Am. Ceram. Soc.* 2003;86:1440-1442.
[450] del Real RP, Padilla S, Vallet-Regi M. Gentamicin release from hydroxyapatite/poly(ethyl methacrylate)/poly(methyl methacrylate)composites. *J. Biomed. Mater. Res.* 2000;52:1-7.
[451] Saito M, Maruoka A, Mori T, Sugano N, Hino K. Experimental studies on a new bioactive bone cement: hydroxyapatite composite resin. *Biomaterials* 1994;15:156-160.
[452] Watson KE, Ten Huisen KS, Brown PW. The formation of hydroxyapatite – calcium polyacrylate composites. *J. Mater. Sci. Mater. Med.* 1999;10:205-213.
[453] Reed CS, Ten Huisen KS, Brown PW, Allcock HR. Thermal stability and compressive strength of calcium-deficient hydroxyapatite – poly[bis(carboxylatophenoxy)phosphazene] composites. *Chem. Mater.* 1996;8:440-447.
[454] Peter SJ, Kim P, Yasko AW, Yaszemski MJ, Mikos AG. Crosslinking characteristics of an injectable poly(propylene fumarate)/β-tricalcium phosphate paste and mechanical properties of the crosslinked composite for use as a biodegradable bone cement. *J. Biomed. Mater. Res.* 1999;44:314-321.
[455] He S, Yaszemski MJ, Yasko AW, Engel PS, Mikos AG. Injectable biodegradable polymer composites based on poly (propylene fumarate) crosslinked with poly(ethylene glycol)-dimethacrylate. *Biomaterials* 2000;21:2389-2394.
[456] Ignjatovic N, Jovanovic J, Suljovrujic E, Uskokovic D. Injectable polydimethylsiloxane/hydroxyapatite composite cement. *Biomed. Mater. Eng.* 2003;13:401-410.
[457] Fujishiro Y, Takahashi K, Sato T. Preparation and compressive strength of α-tricalcium phosphate/gelatin gel composite cement. *J. Biomed. Mater. Res.* 2001;54:525-230.
[458] Miyazaki K, Horibe T, Antonucci JM, Takagi S, Chow LC. Polymeric calcium phosphate cements: analysis of reaction products and properties. *Dental Mater.* 1993;9:41-45.
[459] Miyazaki K, Horibe T, Antonucci JM, Takagi S, Chow LC. Polymeric calcium phosphate cements: setting reaction modifiers. *Dental Mater.* 1993;9:46-50.
[460] Dos Santos LA, De Oliveira LC, Rigo ECS, Carrodeguas RG, Boschi AO, De Arruda ACF. Influence of polymeric additives on the

mechanical properties of α-tricalcium phosphate cement. *Bone* 1999;25:99S-102S.

[461] Greish YE, Brown PW, Bender JD, Allcock HR, Lakshmi S, Laurencin CT. Hydroxyapatite-polyphosphazane composites prepared at low temperatures. *J. Am. Ceram. Soc.* 2007;90:2728-2734.

[462] Greish YE, Bender JD, Lakshmi S, Brown PW, Allcock HR, Laurencin CT. Formation of hydroxyapatite-polyphosphazene polymer composites at physiologic temperature. *J. Biomed. Mater. Res. A* 2006;77A:416-425.

[463] Greish YE, Bender JD, Lakshmi S, Brown PW, Allcock HR, Laurencin CT. Low temperature formation of hydroxyapatite-poly(alkyl oxybenzoate) phosphazene composites for biomedical applications. *Biomaterials* 2005;26:1-9.

[464] Mickiewicz RA, Mayes AM, Knaack D. Polymer – calcium phosphate cement composites for bone substitutes. *J. Biomed. Mater. Res.* 2002;61:581-592.

[465] Carey LE, Xu HHK, Simon CG, Takagi S, Chow LC. Premixed rapid-setting calcium phosphate composites for bone repair. *Biomaterials* 2005;26:5002-5014.

[466] Miao X, Tan LP, Tan LS, Huang X. Porous calcium phosphate ceramics modified with PLGA–bioactive glass. *Mater. Sci. Eng. C* 2007;27:274-279.

[467] Lickorish D, Guan L, Davies JE. A three-phase, fully resorbable, polyester/calcium phosphate scaffold for bone tissue engineering: evolution of scaffold design. *Biomaterials* 2007;28:1495-1502.

[468] Xu HHK, Simon CG. Fast setting calcium phosphate-chitosan scaffold: mechanical properties and biocompatibility. *Biomaterials* 2005;26:1337-1348.

[469] Zhang L, Li Y, Zhou G, Lu GY, Zuo Y. Setting mechanism of nano-hydroxyapatite/chitosan bone cement. *J. Inorg. Mater.* 2006;21:1197-1202.

[470] Ruhe PQ, Hedberg EL, Padron NT, Spauwen PHM, Jansen JA, Mikos AG. Biocompatibility and degradation of poly(DL-lactic-*co*-glycolic acid)/calcium phosphate cement composites. *J. Biomed. Mater. Res. A* 2005;74A:533-544.

[471] Guo DG, Sun HL, Xu KW, Han Y. Long-term variations in mechanical properties and *in vivo* degradability of CPC/PLGA composite. *J. Biomed. Mater. Res. B Appl. Biomater.* 2007;82B:533-544.

[472] Habraken WJEM, Wolke JGC, Mikos AG, Jansen JA. Injectable PLGA microsphere/calcium phosphate cements: physical properties and

degradation characteristics. *J. Biomater. Sci. Polym. Edn.* 2006;17:1057-1074.
[473] Ruhe PQ, Hedberg-Dirk EL, Padron NT, Spauwen PHM, Jansen JA, Mikos AG. Porous poly (DL-lactic-*co*-glycolic acid)/calcium phosphate cement composite for reconstruction of bone defects. *Tissue Eng.* 2006;12:789-800.
[474] Ruhe PQ, Hedberg EL, Padron NT, Spauwen PHM, Jansen JA, Mikos AG. rhBMP-2 release from injectable poly (DL-lactic-*co*-glycolic acid)/calcium phosphate composites. *J. Bone Joint Surg.* (Am.) 2003;85A(suppl 3):75-81.
[475] Ruhe PQ, Boerman OC, Russel FGM, Spauwen PHM, Mikos AG, Jansen JA. Controlled release of rhBMP-2 loaded poly (DL-lactic-*co*-glycolic acid)/calcium phosphate cement composites *in vivo*. *J. Control. Release* 2005;106:162-171.
[476] Plachokova A, Link D, van den Dolder J, van den Beucken J, Jansen JA. Bone regenerative properties of injectable PGLA-CaP composite with TGF-β1 in a rat augmentation model. *J. Tissue. Eng. Regen. Med.* 2007;1:457-464.
[477] Webster TJ, Siegel RW, Bizios R. Osteoblast adhesion on nanophase ceramics. *Biomaterials* 1999;20:1221-1227.
[478] Webster TJ, Ergun C, Doremus RH, Siegel RW, Bizios R. Specific proteins mediate enhanced osteoblast adhesion on nanophase ceramics. *J. Biomat. Med. Res.* 2000;51:475-483.
[479] Webster TJ, Ergun C, Doremus RH, Siegel RW, Bizios R. Enhanced functions of osteoblasts on nanophase ceramics. *Biomaterials* 2000;21:1803-1810.
[480] Li G, Huang J, Li Y, Zhang R, Deng B, Zhang J, Aoki H. *In vitro* study on influence of a discrete nano-hydroxyapatite on leukemia P388 cell behavior. *Biomed. Mater. Eng.* 2007;17:321-327.
[481] Tadic D, Peters F, Epple M. Continuous synthesis of amorphous carbonated apatites. *Biomaterials* 2002;23:2553-2559.
[482] Xu HHK, Sun L, Weir MD, Antonucci JM, Takagi S, Chow LC, Peltz M. Nano DCPA-whisker composites with high strength and Ca and PO_4 release. *J. Dent. Res.* 2006;85:722-727.
[483] Xu HHK, Weir MD, Sun L, Takagi S, Chow LC. Effects of calcium phosphate nanoparticles on Ca-PO_4 composite. *J. Dent. Res.* 2007;86:378-383.

[484] Xu HHK, Weir MD, Sun L. Nanocomposites with Ca and PO_4 release: effects of reinforcement, dicalcium phosphate particle size and silanization. *Dent. Mater.* 2007;23:1482-1491.

[485] Xu HHK, Sun L, Weir MD, Takagi S, Chow LC, Hockey B. Effects of incorporating nanosized calcium phosphate particles on properties of whisker-reinforced dental composites. *J. Biomed. Mater. Res. B Appl. Biomater.* 2007;81B:116-125.

[486] Deng XM, Hao JY, Wang CS. Preparation and mechanical properties of nanocomposites of poly(D,L-lactide) with Ca-deficient hydroxyapatite nanocrystals. *Biomaterials* 2001;22:2867-2873.

[487] Hong ZK, Zhang PB, He CL, Qiu XY, Liu AX, Chen L, Chen X, Jing X. Nanocomposite of poly(L-lactide) and surface grafted hydroxyapatite: mechanical properties and biocompatibility. *Biomaterials* 2005;26:6296-6304.

[488] Deng C, Weng J, Cheng QY, Zhou SB, Lu X, Wan JX, Qu SX, Feng B, Li XH. Choice of dispersants for the nano-apatite filler of polylactide-matrix composite biomaterial. *Current Applied Physics* 2007;7:679-682.

[489] Deng C, Weng J, Lu X, Zhou SB, Wan JX, Qu SX, Feng B, Li XH, Cheng QY. Mechanism of ultrahigh elongation rate of poly(D,L-lactide)-matrix composite biomaterial containing nano-apatite fillers. *Mater. Lett.* 2008;62:607-610.

[490] Kothapalli CR, Shaw MT, Wei M. Biodegradable HA-PLA 3-D porous scaffolds: effect of nano-sized filler content on scaffold properties. *Acta Biomater.* 2005;1:653-662.

[491] Hong Z, Qiu X, Sun J, Deng M, Chen X, Jing X. Grafting polymerization of L-lactide on the surface of hydroxyapatite nano-crystals. *Polymer* 2004;45:6699-6706.

[492] Xiao Y, Li D, Fan H, Li X, Gu Z, Zhang X. Preparation of nano-HA/PLA composite by modified-PLA for controlling the growth of HA crystals. *Mater. Lett.* 2007;61:59-62.

[493] Qiu X, Han Y, Zhuang X, Chen X, Li Y, Jing X. Preparation of nano-hydroxyapatite/poly(L-lactide) biocomposite microspheres. *J. Nanoparticle Res.* 2007;9:901-908.

[494] Kim SS, Park MS, Jeon Q, Choi CY, Kim BS. Poly(lactide-*co*-glycolide)/hydroxyapatite composite scaffolds for bone tissue engineering. *Biomaterials* 2006;27:1399-1409.

[495] Hong Z, Zhang P, Liu A, Chen L, Chen X, Jing X. Composites of poly(lactide-*co*-glycolide) and the surface modified carbonated

hydroxyapatite nanoparticles. *J. Biomed. Mater. Res. A* 2007;81A:515-522.
[496] Huang YX, Ren J, Chen C, Ren TB, Zhou XY. Preparation and properties of poly(lactide-*co*-glycolide) (PLGA)/nano-hydroxyapatite (NHA) scaffolds by thermally induced phase separation and rabbit MSCs culture on scaffolds. *J. Biomaterials Applications* 2008;22:409-432.
[497] Du C, Cui FZ, Zhu XD, de Groot K. Three-dimensional nano-HAp/collagen matrix loading with osteogenic cells in organ culture. *J. Biomed. Mater. Res.* 1999;44:407-415.
[498] Wang RZ, Cui FZ, Lu HB, Wen HB, Ma CL, Li HD. Synthesis of nanophase hydroxyapatite/collagen composite. *J. Mater. Sci. Lett.* 1995;14:490-492.
[499] Du C, Cui FZ, Feng QL, Zhu XD, de Groot K. Tissue response to nano-hydroxyapatite/collagen composite implants in marrow cavity. *J. Biomed. Mater. Res.* 1998;42:540-548.
[500] Kikuchi M, Itoh S, Ichinose S, Shinomiya K, Tanaka J. Self-organization mechanism in a bone-like hydroxyapatite/collagen nanocomposite synthesized *in vitro* and its biological reaction *in vivo*. *Biomaterials* 2001;22:1705-1711.
[501] Kikuchi M, Matsumoto HN, Yamada T, Koyama Y, Takakuda K, Tanaka J. Glutaraldehyde cross-linked hydroxyapatite/collagen self-organized nanocomposites. *Biomaterials* 2004;25:63-69.
[502] Lynn AK, Nakamura T, Patel N, Porter AE, Renouf AC, Laity PR, Best SM, Cameron RE, Shimizu Y, Bonfield W. Composition-controlled nanocomposites of apatite and collagen incorporating silicon as osseopromotive agent. *J. Biomed. Mater. Res. A* 2005;74A:447-453.
[503] Chang MC, Tanaka J. FTIR study for hydroxyapatite/collagen nanocomposite cross-linked by glutaraldehyde. *Biomaterials* 2002;23:4811-4818.
[504] Chang MC, Tanaka J. XPS study for the microstructure development of hydroxyapatite-collagen nanocomposites cross-linked using glutaraldehyde. *Biomaterials* 2002;23:3879-3885.
[505] Murugan R, Ramakrishna S. In situ formation of recombinant humanlike collagen-hydroxyapatite nanohybrid through bionic approach. *Appl. Phys. Lett.* 2006;88:193124.
[506] Wang Y, Yang C, Chen X, Zhao N. Biomimetic formation of hydroxyapatite/collagen matrix composite. *Adv. Eng. Mater.* 2006;8:97-100.

[507] Thomas V, Dean DR, Jose MV, Mathew B, Chowdhury S, Vohra YK. Nanostructured biocomposite scaffolds based on collagen coelectrospun with nanohydroxyapatite. *Biomacromolecules* 2007;8:631-637.

[508] Fukui N, Sato T, Kuboki Y, Aoki H. Bone tissue reaction of nano-hydroxyapatite/collagen composite at the early stage of implantation. *Biomed. Mater. Eng.* 2008;18:25-33.

[509] Liao SS, Tamura K, Zhu Y, Wang W, Uo M, Akasaka T, Cui FZ, Watari F. Human neutrophils reaction to the biodegraded nano-hydroxyapatite/collagen and nano-hydroxyapatite/collagen/poly(L-lactic acid) composites. *J. Biomed. Mater. Res. A* 2006;76A:820-825.

[510] Liao SS, Cui FZ, Zhu Y. Osteoblasts adherence and migration through three-dimensional porous mineralized collagen based composite: nHAC/PLA. *J. Bioact. Compat. Polym.* 2004;19:117-130.

[511] Liao SS, Cui FZ, Zhang W, Feng QL. Hierarchically biomimetic bone scaffold materials: nano-HA/collagen/PLA composite. *J. Biomed. Mater. Res. B Appl. Biomater.* 2004;69B:158-165.

[512] Liao SS, Cui FZ. *In vitro* and *in vivo* degradation of the mineralized collagen based composite scaffold: nanohydroxyapatite/collagen/poly(L-lactide). *Tissue Eng.* 2004;10:73-80.

[513] Liao SS, Wang W, Uo M, Ohkawa S, Akasaka T, Tamura K, Cui FZ, Watari F. A three-layered nano-carbonated hydroxyapatite/collagen/ PLGA composite membrane for guided tissue regeneration. *Biomaterials* 2005;26:7564-7571.

[514] Li X, Feng Q, Cui FZ. *In vitro* degradation of porous nano-hydroxyapatite/collagen/PLLA scaffold reinforced by chitin fibres. *Mater. Sci. Eng. C* 2006;26:716-720.

[515] Zhou DS, Zhao KB, Li Y, Cui FZ, Lee IS. Repair of segmental defects with nano-hydroxyapatite/collagen/PLA composite combined with mesenchymal stem cells. *J. Bioactive and Compatible Polymers* 2006;21:373-384.

[516] Zhang C, Hu YY, Cui FZ, Zhang SM, Ruan DK. A study on a tissue-engineered bone using rhBMP-2 induced periosteal cells with a porous nano-hydroxyapatite/collagen/poly(L-lactic acid) scaffold. *Biomed. Mater.* 2006;1:56-62.

[517] Liao S, Watari F, Zhu Y, Uo M, Akasaka T, Wang W, Xu G, Cui F. The degradation of the three layered nano-carbonated hydroxyapatite/collagen/PLGA composite membrane *in vitro*. *Dental Materials* 2007;23:1120-1128.

[518] Degirmenbasi N, Kalyon DM, Birinci E. Biocomposites of nanohydroxyapatite with collagen and poly(vinyl alcohol). *Colloids Surf. B Biointerfaces* 2006;48:42-49.
[519] Zhang SM, Cui FZ, Liao SS, Zhu Y, Han L. Synthesis and biocompatibility of porous nanohydroxyapatite/collagen/alginate composite. *J. Mater. Sci. Mater. Med.* 2003;14:641-645.
[520] Sotome S, Uemura T, Kikuchi M, Chen J, Itoh S, Tanaka J, Tateishi T, Shinomiya K. Synthesis and *in vivo* evaluation of a novel hydroxyapatite/collagen-alginate as a bone filler and a drug delivery carrier of bone morphogenetic protein. *Mater. Sci. Eng. C* 2004;24:341-347.
[521] Chang MC, Ko CC, Douglas WH. Conformational change of hydroxyapatite/gelatin nanocomposite by glutaraldehyde. *Biomaterials* 2003;24:3087-3094.
[522] Kim HW, Kim HE, Vehid S. Stimulation of osteoblast responses to biomimetic nanocomposites of gelatin-hydroxyapatite for tissue engineering scaffolds. *Biomaterials* 2005;26:5221-5230.
[523] Chang MC, Ikoma T, Tanaka J. Cross-linkage of hydroxyapatite/gelatin nanocomposite using EGDE. *J. Mater. Sci.* 2004;39:5547-5550.
[524] Teng S, Shi J, Peng B, Chen L. The effect of alginate addition on the structure and morphology of hydroxyapatite/gelatin nanocomposites. *Compos. Sci. Technol.* 2006;66:1532-1538.
[525] Chang MC, Ko CC, Douglas WH. Preparation of hydroxyapatite-gelatin nanocomposite. *Biomaterials* 2003;24:2853-2862.
[526] Mobini S, Javadpour J, Hosseinalipour M, Ghazi-Khansari M, Khavandi A, Rezaie HR. Synthesis and characterisation of gelatin-nano hydroxyapatite composite scaffolds for bone tissue engineering. *Adv. Appl. Ceram.* 2008;107:4-8.
[527] Wang XJ, Li Y, Wei J, de Groot K. Development of biomimetic nano-hydroxyapatite/poly(hexamethylene adipamide) composites. *Biomaterials* 2002;23:4787-4791.
[528] Lewandrowski KU, Bondre SP, Wise DL, Trantolo DJ. Enhanced bioactivity of a poly(propylene fumarate) bone graft substitute by augmentation with nano-hydroxyapatite. *Biomed. Mater. Eng.* 2003;13:115-124.
[529] Wei J, Li Y, He Y. Processing properties of nano apatite-polyamide biocomposites. *J. Mater. Sci.* 2005;40:793-796.
[530] Wei J, Li Y, Chen W, Zuo Y. A study on nano-composite of hydroxyapatite and polyamide. *J. Mater. Sci.* 2003;38:3303-3306.

[531] Wei J, Li Y. Tissue engineering scaffold material of nano-apatite crystals and polyamide composite. *Eur. Polym. J.* 2004;40:509-515.
[532] Wang H, Li Y, Zuo Y, Li J, Ma S, Cheng L. Biocompatibility and osteogenesis of biomimetic nano-hydroxyapatite/polyamide composite scaffolds for bone tissue engineering. *Biomaterials* 2007;28:3338-3348.
[533] Sender C, Dantras E, Dantras-Laffont L, Lacoste MH, Dandurand J, Mauzac M, Lacout JL, Lavergne C, Demont P, Bernes A, Lacabanne C. Dynamic mechanical properties of a biomimetic hydroxyapatite/polyamide 6,9 nanocomposite. *J. Biomed. Mater. Res. B Appl. Biomater.* 2007;83B:628-635.
[534] Yang K, Wei J, Wang CY, Li Y. A study on *in vitro* and *in vivo* bioactivity of nano hydroxyapatite/polymer biocomposite. *Chinese Science Bulletin* 2007;52:267-271.
[535] Zhang X, Li Y, Lv GY, Zuo Y, Mu YH. Thermal and crystallization studies of nano-hydroxyapatite reinforced polyamide 66 biocomposites. *Polym. Degrad. Stabil.* 2006;91:1202-1207.
[536] Huang M, Feng J, Wang J, Zhang X, Li Y. Yan Y. Synthesis and characterization of nano-HA/PA66 composite. *J. Mater. Sci. Mater. Med.* 2003;14:655-660.
[537] Zhang X, Li Y, Zuo Y, Lv GY, Mu YH, Li H. Morphology, hydrogen-bonding and crystallinity of nano-hydroxyapatite/polyamide 66 biocomposites. *Composites* A 2007;38:843-848.
[538] Lan W, Li Y, Yi Z, Li Z, Mu YH, Jimei H. Study on the biomimetic properties of bone substitute material: nano-hydroxyapatite/polyamide 66 composite. *Mater. Sci. Forum* 2006;510-511:938-941.
[539] Zhang L, Li Y, Wang X, Wei J, Peng X. Studies on the porous scaffold made of the nano-HA/PA66 composite. J. Mater. Sci. 2005;40:107-110.
[540] Zhang X, Li Y, Lv GY, Zuo Y, Mu YH, Lan W. The study on interaction mechanism between n-HA and PA66 in n-HA/PA66 biocomposites. *Funct. Mater.* 2005;36:896-899.
[541] Yusong P, Dangsheng X, Xiaolin C. Mechanical properties of nanohydroxyapatite reinforced poly(vinyl alcohol) gel composites as biomaterial. *J. Mater. Sci.* 2007;42:5129-5134.
[542] Xu F, Li Y, Wang X, Wei J, Yang A. Preparation and characterization of nano-hydroxyapatite/poly(vinyl alcohol) hydrogel biocomposite. *J. Mater. Sci.* 2004;39:5669-5672.
[543] Wang HS, Wang GX, Pan QX. Electrochemical study of the interactions of DNA with redox-active molecules based on the immobilization of

dsDNA on the sol-gel derived nano porous hydroxyapatite-polyvinyl alcohol hybrid material coating. *Electroanalysis* 2005;17:1854-1860.

[544] Pramanik N, Mohapatra S, Pramanik P, Bhargava P. Processing and properties of nano-hydroxyapatite (n-HAp)/poly(ethylene-*co*-acrylic acid) (EAA) composite using a phosphonic acid coupling agent for orthopedic applications. *J. Am. Ceram. Soc.* 2007;90:369-375.

[545] Pramanik N, Bhargava P, Alam S, Pramanik P. Processing and properties of nano- and macro-hydroxyapatite/poly(ethylene-*co*-acrylic acid) composites. *Polym. Compos.* 2006;27:633-641.

[546] Zhang L, Li Y, Yang A, Peng X, Wang X, Zhang X. Preparation and *in vitro* investigation of chitosan/nano-hydroxyapatite composite used as bone substitute materials. *J. Mater. Sci. Mater. Med.* 2005;16:213-219.

[547] Zhang YF, Cheng XR, Chen Y, Shi B, Chen XH, Xu DX, Ke J. Three-dimensional nanohydroxyapatite/chitosan scaffolds as potential tissue engineered periodontal tissue. *J. Biomaterials Applications* 2007;21:333-349.

[548] Kong L, Gao Y, Lu G, Gong Y, Zhao N, Zhang X. A study on the bioactivity of chitosan/nano-hydroxyapatite composite scaffolds for bone tissue engineering. *Eur. Polymer J.* 2006;42:3171-3179.

[549] Lu XY, Wang XH, Qu SX, Weng J. Preparation of nano-hydroxyapatite/chitosan hybrids. *J. Inorg. Mater.* 2008;23:332-336.

[550] Zhou G, Li Y, Zhang L, Li H, Wang M, Cheng L, Wang Y, Wang H, Shi P. The study of tri-phasic interactions in nano-hydroxyapatite/konjac glucomannan/chitosan composite. *J. Mater. Sci.* 2007;42:2591-2597.

[551] Huang J, Lin YW, Fu XW, Best SM, Brooks RA, Rushton N, Bonfield W. Development of nano-sized hydroxyapatite reinforced composites for tissue engineering scaffolds. *J. Mater. Sci. Mater. Med.* 2007;18:2151-2157.

[552] Lee HJ, Choi HW, Kim KJ, Lee SC. Modification of hydroxyapatite nanosurfaces for enhanced colloidal stability and improved interfacial adhesion in nanocomposites. *Chem. Mater.* 2006;18:5111-5118.

[553] Lee HJ, Kim SE, Choi HW, Kim CW, Kim KJ, Lee SC. The effect of surface-modified nano-hydroxyapatite on biocompatibility of poly(ε-caprolactone)/hydroxyapatite nanocomposites. *Eur. Polymer J.* 2007;43:1602-1608.

[554] Pang P, Li W, Liu Y. Effect of ball milling process on the microstructure of titanium-nanohydroxyapatite composite powder. *Rare Metals* 2007;26:118-123.

[555] Li W, Pang P, Liu Y. Microstructure and phase composition of Ti-based biocomposites with different contents of nano-hydroxyapatite. *Transactions of Nonferrous Metals Society of China* 2007;17:Special Issue:S1148-S1151.
[556] Hao JY, Liu Y, Zhou S, Li Z, Deng X. Investigation of nanocomposites based on semi-interpenetrating network of [L-poly (ε-caprolactone)]/[net-poly (ε-caprolactone)] and hydroxyapatite nanocrystals. *Biomaterials* 2003;24:1531-1539.
[557] Yan Y, Li Y, Zheng Y, Yi Z, Wei J, Xia C, Chen Y. Synthesis and properties of a copolymer of poly(1,4-phenylene sulfide)-poly(2,4-phenylene sulfide acid) and its nano-apatite reinforced composite. *Eur. Polym. J.* 2003;39:411-416.
[558] Bhattacharyya S, Nair LS, Singh A, Krogman NR, Bender J, Greish YE, Brown PW, Allcock HR, Laurencin CT. Development of biodegradable polyphosphazene-nanohydroxyapatite composite nanofibers via electrospinning. *MRS Symp. Proc.* 2005;845:91-96.
[559] Sinha A, Nayar S, Agrawal A, Bhattacharyya D, Ramachandrarao P. Synthesis of nanosized and microporous precipitated hydroxyapatite in synthetic polymers and biopolymers. *J. Am. Ceram. Soc.* 2003;86:357-359.
[560] Zuo Y, Li Y, Wei J, Han J, Xu F. The preparation and characterization of n-HA/PA series biomedical composite. *Funct. Mater.* 2004;35:513-516.
[561] Zhou G, Li Y, Zhang L, Zuo Y, Jansen JA. Preparation and characterization of nano-hydroxyapatite/chitosan/konjac glucomannan composite. *J. Biomed. Mater. Res. A* 2007;83A:931-939.
[562] Daniel-da-Silva AL, Lopes AB, Gil AM, Correia RN. Synthesis and characterization of porous κ-carrageenan/calcium phosphate nanocomposite scaffolds. *J. Mater. Sci.* 2007;42:8581-8591.
[563] Furuzono T, Kishida A, Tanaka J. Nano-scaled hydroxyapatite/polymer composite I. Coating of sintered hydroxyapatite particles on poly(γ-methacryloxypropyl trimethoxysilane)-grafted silk fibroin fibers through chemical bonding. *J. Mater. Sci. Mater. Med.* 2004;15:19-23.
[564] Korematsu A, Furuzono T, Yasuda S, Tanaka J, Kishida A. Nano-scaled hydroxyapatite/polymer composite II. Coating of sintered hydroxyapatite particles on poly(2-(o-[1'-methylpropylideneamino] carboxyamino) ethyl methacrylate)-grafted silk fibroin fibers through covalent linkage. *J. Mater. Sci.* 2004;39:3221-3225.

[565] Korematsu A, Furuzono T, Yasuda S, Tanaka J, Kishida A. Nano-scaled hydroxyapatite/polymer composite III. Coating of sintered hydroxyapatite particles on poly(4-methacryloyloxyethyl trimellitate anhydride)-grafted silk fibroin fibers. *J. Mater. Sci. Mater. Med.* 2005;16:67-71.
[566] Yang K, Wang C, Wei J. A study on biocomposite of nano apatite/poly (1,4-phenylene-sulfide)-poly (2,4-phenylene sulfide acid). *Composites B* 2007;38:306-310.
[567] Jiang L, Li Y, Zhang L, Wang XJ. Study on nano-hydroxyapatite/chitosan-carboxymethyl cellulose composite scaffold. *J. Inorg. Mater.* 2008;23:135-140.
[568] Jiang L, Li Y, Zhang L, Liao J. Preparation and properties of a novel bone repair composite: nano-hydroxyapatite/chitosan/carboxymethyl cellulose. *J. Mater. Sci. Mater. Med.* 2008;19:981-987.
[569] Liou SC, Chen SY, Liu DM. Phase development and structural characterization of calcium phosphate ceramics-polyacrylic acid nanocomposites at room temperature in water-methanol mixtures. *J. Mater. Sci. Mater. Med.* 2004;15:1261-1266.
[570] Liu L, Liu J, Wang M, Min S, Cai Y, Zhu L, Yao J. Preparation and characterization of nano-hydroxyapatite/silk fibroin porous scaffolds. *J. Biomater. Sci. Polymer Edn.* 2008;19:325-338.
[571] Ren YJ, Sun XD, Cui FZ, Wei YT, Cheng ZJ, Kong XD. Preparation and characterization of *antheraea pernyi* silk fibroin based nanohydroxyapatite composites. *J Bioactive and Compatible Polymers* 2007;22:465-474.
[572] Mikołajczyk T, Rabiej S, Bogun M. Analysis of the structural parameters of polyacrylonitrile fibers containing nanohydroxyapatite. *J. Appl. Polym. Sci.* 2006;101:760-765.
[573] Wei J, Li Y, Lau KT. Preparation and characterization of a nano apatite/polyamide$_6$ bioactive composite. *Composites* B 2007;38:301-305.
[574] Sundaraseelan J, Sastry TP. Fabrication of a biomimetic compound containing nano hydroxyapatite – demineralised bone matrix. *J. Biomed. Nanotechnol.* 2007;3:401-405.
[575] Leeuwenburgh SCG, Jansen JA, Mikos AG. Functionalization of oligo(poly(ethylene glycol)fumarate) hydrogels with finely dispersed calcium phosphate nanocrystals for bone-substituting purposes. *J. Biomater. Sci. Polymer Edn.* 2007;18:1547-1564.
[576] Sun TS, Guan K, Shi SS, Zhu B, Zheng YJ, Cui FZ, Zhang W, Liao SS. Effect of nano-hydroxyapatite/collagen composite and bone

morphogenetic protein-2 on lumbar intertransverse fusion in rabbits. *Chin. J. Traumatol.* 2004;7:18-24.

[577] Itoh S, Kikuehi M, Koyama Y, Takakuda K, Shinomiya K, Tanaka J. Development of a hydroxyapatite/collagen nanocomposite as a medical device. *Cell Transp.* 2004;13:451-461.

[578] Hu Q, Li BQ, Wang M, Shen JC. Preparation and characterization of biodegradable chitosan/hydroxyapatite nanocomposite rods via *in situ* hybridization: a potential material as internal fixation of bone fracture. *Biomaterials* 2004;25:779-785.

[579] Wei G, Ma PX. Structure and properties of nano-hydroxyapatite/polymer composite scaffolds for bone tissue engineering. *Biomaterials* 2004;25:4749-4757.

[580] Liou SC, Chen SY, Liu DM. Synthesis and characterization of needlelike apatitic nanocomposite with controlled aspect ratios. *Biomaterials* 2003;24:3981-3988.

[581] Liou SC, Chen SY, Liu DM. Manipulation of nanoneedle and nanosphere apatite/poly(acrylic acid) nanocomposites. *J. Biomed. Mater. Res. B Appl. Biomater.* 2005;73B:117-122.

[582] Huang J, Best SM, Bonfield W, Brooks RA, Rushton N, Jayasinghe SN, Edirisinghe MJ. *In vitro* assessment of the biological response to nano-size hydroxyapatite. *J. Mater. Sci. Mater. Med.* 2004;15:441-415.

[583] Kong L, Gao Y, Cao W, Gong Y, Zhao N, Zhang X. Preparation and characterization of nano-hydroxyapatite/chitosan composite scaffolds. *J. Biomed. Mater. Res. A* 2005;75A:275-282.

[584] Christenson EM, Anseth KS, van den Beucken JJJP, Chan CK, Ercan B, Jansen JA, Laurencin CT, Li WJ, Murugan R, Nair LS, Ramakrishna S, Tuan RS, Webster TJ, Mikos AG. Nanobiomaterial applications in orthopedics. *J. Orthop. Res.* 2007;25:11-22.

[585] Nimni ME. (Ed.), *Collagen.* CRC Press, Boca Raton, FL, 1988.

[586] Olmo N, Turnay J, Herrera JI, Gavilanes JG, Lizarbe MA. Kinetics of *in vivo* degradation of sepiolite-collagen complexes: Effect of glutaraldehyde treatment. *J. Biomed. Mater. Res.* 1996;30:77-84.

[587] Xie J, Baumann MJ, McCabe LR. Osteoblasts respond to hydroxyapatite surfaces with immediate changes in gene expression. *J. Biomed. Mater. Res. A* 2004;71A:108-117.

[588] Tcacencu I, Wendel M. Collagen-hydroxyapatite composite enhances regeneration of calvaria bone defects in young rats but postpones the regeneration of calvaria bone in aged rats. *J. Mater. Sci. Mater. Med.* 2008;19:2015-2021.

[589] Yamauchi K, Goda T, Takeuchi N, Einaga H, Tanabe T. Preparation of collagen/calcium phosphate multilayer sheet using enzymatic mineralization. *Biomaterials* 2004;25:5481-5489.
[590] Du C, Cui FZ, Zhang W, Feng QL, Zhu XD, de Groot K. Formation of calcium phosphate/collagen composites through mineralization of collagen matrix. *J. Biomed. Mater. Res.* 2000;50:518-527.
[591] Hellmich C, Ulm FJ. Are mineralized tissues open crystal foams reinforced by crosslinked collagen? – some energy arguments. *J. Biomech.* 2002;35:1199-1212.
[592] Boskey AL. Hydroxyapatite formation in a dynamic collagen gel system – effects of type I collagen, lipids and proteoglycans. *J. Phys. Chem.* 1989;93:1628-1633.
[593] Mathers NJ, Czernuszka JT. Growth of hydroxyapatite on type I collagen. *J. Mater. Sci. Lett.* 1991;10:992-993.
[594] Sukhodub LF, Moseke C, Sukhodub LB, Sulkio-Cleff B, Maleev VYa, Semenov MA, Bereznyak EG, Bolbukh TV. Collagen-hydroxyapatite-water interactions investigated by XRD, piezogravimetry, infrared and Raman spectroscopy. *J. Molecular Structure* 2004;704:53-58.
[595] Roveri N, Falini G, Sidoti MC, Tampieri A, Landi E, Sandri M, Parma B. Biologically inspired growth of hydroxyapatite nanocrystals inside self-assembled collagen fibers. *Mater. Sci. Eng. C* 2003;23:441-446.
[596] Tampieri A, Celotti G, Landi E. From biomimetic apatites to biologically inspired composites. *Anal. Bioanal. Chem.* 2005;381:568-576.
[597] Tampieri A, Celotti G, Landi E, Sandri M, Roveri N, Falini G. Biologically inspired synthesis of bone-like composite: self-assembled collagen fibers/hydroxyapatite nanocrystals. *J. Biomed. Mater. Res. A* 2003;67A:618-625.
[598] Mehlisch DR, Taylor TD, Leibold DG, Hiatt R, Waite DE, Waite PD, Laskin DM, Smith ST, Koretz MM. Evaluation of collagen/hydroxyapatite for augmenting deficient alveolar ridges, a preliminary report. *J. Oral Maxillofac. Surg.* 1987;45:408-413.
[599] Okazaki M, Ohmae H, Takahashi J, Kimura H, Sakuda M. Insolubilized properties of UV-irradiated CO_3 apatite-collagen composites. *Biomaterials* 1990;11:568-572.
[600] Ten Huisen KS, Martin RI, Klimkiewicz M, Brown PW. Formation and properties of a synthetic bone composite: hydroxyapatite-collagen. *J. Biomed. Mater. Res.* 1995;29:803-810.

[601] Marouf HA, Quayle AA, Sloan P. *In vitro* and *in vivo* studies with collagen/hydroxyapatite implants. *Int. J. Oral Maxillofac. Implants* 1990;5:148-154.
[602] Zerwekh JE, Kourosh S, Scheinberg R, Kitano T, Edwards ML, Shin D, Selby DK. Fibrillar collagen-biphasic calcium phosphate composite as a bone graft substitute for spinal fusion. *J. Orthop. Res.* 1992;10:565-572.
[603] Clarke KI, Graves SE, Wong ATC, Triffit JT, Francis MJO, Czernuszka JT. Investigation into the formation and mechanical properties of a bioactive material based on collagen and calcium phosphate. *J. Mater. Sci. Mater. Med.* 1993;4:107-110.
[604] Rovira A, Bareille R, Lopez L, Rouasis F, Bordenave L, Rey C, Rabaud M. Preliminary report on a new composite material made of calcium phosphate, elastin peptides and collagens. *J. Mater. Sci. Mater. Med.* 1993;4:372-380.
[605] Zhang QQ, Ren L, Wang C, Liu LR, Wen XJ, Liu YH, Zhang XD. Porous hydroxyapatite reinforced with collagen protein. *Artif. Cells Blood Substit. Immobiliz. Biotechnol.* 1996;24:693-702.
[606] Bakoš D, Soldán M, Hernández-Fuentes I. Hydroxyapatite-collagen-hyaluronic acid composite. *Biomaterials* 1999;20:191-195.
[607] John A, Hong L, Ikada Y, Tabata Y. A trial to prepare biodegradable collagen-hydroxyapatite composites for bone repair. *J. Biomater. Sci. Polym. Edn.* 2001;12:689-705.
[608] ltoh S, Kikuchi M, Takakuda K, Koyama Y, Matsumoto HN, Ichinose S, Tanaka J, Kawauchi T, Shinomiya K. The biocompatibility and osteoconductive activity of a novel hydroxyapatite/collagen composite biomaterial and its function as a carrier of rhBMP-2. *J. Biomed. Mater. Res.* 2001;54:445-453.
[609] Shinomiya K, Itoh S, Kawauchi T, Kikuchi M, Tanaka J. Development of a novel hydroxyapatite/collagen composite biomaterial. *Tissue Eng. Therap. Use* 2001;5:165-177.
[610] Uskokovic V, Ignjatovic N, Petranovic N. Synthesis and characterization of hydroxyapatite-collagen biocomposite materials. *Mater. Sci. Forum* 2002;413:269-274.
[611] Wahl DA, Czernuszka JT. Collagen-hydroxyapatite composites for hard tissue repair. *Eur. Cells Mater.* 2006;11:43-56.
[612] Ishikawa H, Koshino T, Takeuchi R, Saito T. Effects of collagen gel mixed with hydroxyapatite power on interface between newly formed bone and grafted Achilles tendon in rabbit femoral bone tunnel. *Biomaterials* 2001;22:1689-1694.

[613] Sachlos E, Gotora D, Czernuszka JT. Collagen scaffolds reinforced with biomimetic composite nano-sized carbonate-substituted hydroxyapatite crystals and shaped by rapid prototyping to contain internal microchannels. *Tissue Eng.* 2006;12:2479-2487.
[614] Venugopal J, Ramakrishna S, Low S, Choon AT, Kumar TSS. Mineralization of osteoblasts with electrospun collagen/hydroxyapatite nanofibers. *J. Mater. Sci. Mater. Med.* 2008;19:2039-2046.
[615] Teng SH, Lee EJ, Park CS, Choi WY, Shin DS, Kim HE. Bioactive nanocomposite coatings of collagen/hydroxyapatite on titanium substrates. *J. Mater. Sci. Mater. Med.* 2008;19:2453-2461.
[616] Song JH, Kim HE, Kim HW. Collagen-apatite nanocomposite membranes for guided bone regeneration. *J. Biomed. Mater. Res. B Appl. Biomater.* 2007;83B:248-257.
[617] Mittelmeier H, Nizzard M. Knochenregeneration mit industriell gefertigtem Collagen-Apatite implantat. In: Osteogenese und Knochenwachstum (Eds. Hackenbroch MH, Refior HJ, Jäger MG.). *Thieme.* Stuttgart, Germany. 1982.
[618] Serre CM, Papillard M, Chavassieux P, Boivin G. *In vitro* induction of a calcifying matrix by biomaterials constituted of collagen and/or hydroxyapatite: an ultrastructural comparison of three types of biomaterials. *Biomaterials* 1993;14:97-106.
[619] Scabbia A, Trombelli L. A comparative study on the use of a HA/collagen/chondroitin sulphate biomaterial (Biostite&reg) and a bovine-derived HA xenograft (Bio-Oss&reg) in the treatment of deep intraosseous defects. *J. Clin. Periodontol.* 2004;31:348-355.
[620] Yamasaki Y, Yoshida Y, Okazaki M, Shimazu A, Kubo T, Akagawa Y, Uchida T. Action of FGMgCO$_3$Ap-collagen composite in promoting bone formation. *Biomaterials* 2003;24:4913-4920.
[621] Wang X, Grogan SP, Rieser F, Winkelmann V, Maquet V, Berge ML. Mainil-Varlet P. Tissue engineering of biphasic cartilage constructs using various biodegradable scaffolds: an *in vitro* study. *Biomaterials* 2004;25:3681-3688.
[622] Chang MC, Ikonama T, Kikuchi M, Tanaka J. The cross-linkage effect of hydroxyapatite/collagen nanocomposites on a self-organization phenomenon. *J. Mater. Sci. Mater. Med.* 2002;13:993-997.
[623] Iijima M, Moriwaki Y, Kuboki Y. Oriented growth of octacalcium phosphate on and inside the collagenous matrix *in vitro*. *Connect. Tissue Res.* 1996;32:519-524.

[624] Miyamoto Y, Ishikawa K, Takechi M, Toh T, Yuasa T, Nagayama M, Suzuki K. Basic properties of calcium phosphate cement containing atelocollagen in its liquid or powder phases. *Biomaterials* 1998;19:707-715.
[625] Iijima M, Moriwaki Y, Kuboki Y. *In vitro* crystal growth of octacalcium phosphate on type I collagen fiber. *J. Cryst. Growth* 1994;137:553-560.
[626] Iijima M, Iijima K, Moriwaki Y, Kuboki Y. Oriented growth of octacalcium phosphate crystals on type 1 collagen fibrils under physiological conditions. *J. Cryst. Growth* 1994;140:91-99.
[627] Lawson AC, Czernuszka JT. Collagen – calcium phosphate composites. *Proc. Instn. Mech. Engrs. Part H* 1998;212:413-425.
[628] Itoh S, Kikuchi M, Takakuda K, Nagaoka K, Koyama Y, Tanaka J, Shinomiya K. Implantation study of a novel hydroxyapatite/collagen (HAp/col) composite into weight-bearing sites of dogs. *J. Biomed. Mater. Res.* 2002;63:507-515.
[629] Kikuchi M, Ikoma T, Itoh S, Matsumoto HN, Koyama Y, Takakuda K, Shinomiya K, Tanaka J. Biomimetic synthesis of bone-like nanocomposites using the self-organization mechanism of hydroxyapatite and collagen. *Compos. Sci. Technol.* 2004;64:819-825.
[630] Yang XB, Bhataagar RS, Li S, Oreffo RO. Biomimetic collagen scaffolds for human bone cell growth and differentiation. *Tissue Eng.* 2004;14:1148-1159.
[631] Doi Y, Horiguchi T, Moriwaki Y, Kitago H, Kajimoto T, Iwayama Y. Formation of apatite – collagen complexes. *J. Biomed. Mater. Res.* 1996;31:43-49.
[632] Bradt JH, Mertig M, Teresiak A, Pompe W. Biomimetic mineralization of collagen by combined fibril assembly and calcium phosphate formation. *Chem. Mat.* 1999;11:2694-2701.
[633] Scharnweber D, Born R, Flade K, Roessler S, Stoelzel M, Worch H. Mineralization behaviour of collagen type I immobilized on different substrates. *Biomaterials* 2004;25:2371-2380.
[634] Li X, Chang J. Preparation of bone-like apatite-collagen nanocomposites by a biomimetic process with phosphorylated collagen. *J. Biomed. Mater. Res.* A 2008;85A:293-300.
[635] Mai R, Reinstorf A, Pilling E, Hlawitschka M, Jung R, Gelinsky M, Schneider M, Loukota R, Pompe W, Eckelt U, Stadlinger B. Histologic study of incorporation and resorption of a bone cement-collagen composite: an *in vivo* study in the minipig. *Oral Surg. Oral Med. Oral Pathol. Oral Radiol. Endod.* 2008;105:e9-e14.

[636] Young SW, Andrews WA, Muller H, Constantz B. Induction of fracture healing using fibrous calcium phosphate composite spherulites. *Investigative Radiol.* 1991;2:470-473.
[637] Rovira A, Amedee J, Bareille R, Radaud M. Colonization of a calcium phosphate/elastin-solubilized peptide – collagen composite material by human osteoblasts. *Biomaterials*, 1996;17:1535-1540.
[638] Kazim M, Katowitz JA, Fallon M, Piest KL. Evaluation of a collagen/hydroxylapatite implant for cortical bone orbital reconstructive surgery. *Ophthalmic Plastic and Reconstructive Surg.* 1992;8:94-108.
[639] Hirota K, Nishihara K, Tanaka H. Pressure sintering of apatite-collagen composite. *Biomed. Mater. Eng.* 1993;3:147-151.
[640] Zahn D, Hochrein O, Kawska A, Brickmann J, Kniep R. Towards an atomistic understanding of apatite-collagen biomaterials: linking molecular simulation studies of complex-, crystal- and composite-formation to experimental findings. *J. Mater. Sci.* 2007;42:8966-8973.
[641] Silva CC, Pinheiro AG, Figueiro SD, Goes JC, Sasaki JM, Miranda MAR, Sombra ASB. Piezoelectric properties of collagen-nanocrystalline hydroxyapatite composites. *J. Mater. Sci.* 2002;37:2061-2070.
[642] Yunoki S, Ikoma T, Tsuchiya A, Monkawa A, Ohta K, Sotome S, Shinomiya K, Tanaka J. Fabrication and mechanical and tissue ingrowth properties of unidirectionally porous hydroxyapatite/collagen composite. J. Biomed. *Mater. Res. B Appl. Biomater.* 2007;80B:166-173.
[643] Chapman MW, Bucholz R, Cornell C. Treatment of acute fractures with a collagen-calcium phosphate graft material: a randomized clinical trial. *J. Bone Joint Surg. (Am.)* 1997;79A:495-502.
[644] Rodrigues CVM, Serricella P, Linhares ABR, Guerdes RM, Borojevic R, Rossi MA, Duarte MEL, Farina M. Characterization of a bovine collagen-hydroxyapatite composite scaffold for bone tissue engineering. *Biomaterials* 2003;24:4987-4997.
[645] Lickorish D, Ramshaw JAM, Werkmeister JA, Glattauer V, Howlett CR. Development of a collagen-hydroxyapatite composite biomaterial via biomimetic process. *J. Biomed. Mater. Res. A* 2004;68A:19-27.
[646] Hsu FY, Chueh SC, Wang JY. Microspheres of hydroxyapatite/reconstituted collagen as supports for osteoblast cell growth. *Biomaterials* 1999;20:1931-1936.
[647] Wu TJ, Huang HH, Lan CW, Lin CH, Hsu FY, Wang YJ. Studies on the microspheres comprised of reconstituted collagen and hydroxyapatite. *Biomaterials* 2004;25:651-658.

[648] Liao SS, Watari F, Uo M, Ohkawa S, Tamura K, Wang W, Cui FZ. The preparation and characteristics of a carbonated hydroxyapatite/collagen composite at room temperature. *J. Biomed. Mater. Res. B Appl. Biomater.* 2005;74B:817-821.

[649] Yokoyama A, Gelinsky M, Kawasaki T, et al. Biomimetic porous scaffolds with high elasticity made from mineralized collagen – an animal study. *J. Biomed. Mater. Res. B Appl. Biomater.* 2005;75B:464-472.

[650] Zou C, Weng W, Deng XJ, Cheng K, Liu X, Du P, Shen G, Han G. Preparation and characterization of porous β-tricalcium phosphate/collagen composites with an integrated structure. *Biomaterials* 2005;26:5276-5284.

[651] Martins VCA, Goissis G. Nonstoichiometric hydroxyapatite-anionic collagen composite as a support for the double sustained release of gentamicin and norfloxacin/ciprofloxacin. *Artif. Organs* 2000;24:224-230.

[652] Gotterbarm T, Richter W, Jung M, Berardi Vilei S, Mainil-Varlet P, Yamashita T, Breusch SJ. An *in vivo* study of a growth-factor enhanced, cell free, two-layered collagen-tricalcium phosphate in deep osteochondral defects. *Biomaterials* 2006;27:3387-3395.

[653] Martins VC, Goissis G, Ribeiro AC, Marcantonio Jr. E, Bet MR. The controlled release of antibiotic by hydroxyapatite: anionic collagen composites. *Artif. Organs* 1998;22:215-221.

[654] Jayaraman M, Subramanian MV. Preparation and characterization of two new composites: collagen-brushite and collagen-octacalcium phosphate. *Medical Science Monitor* 2002;8:BR481-BR487.

[655] Ikeda H, Yamaza T, Yoshinari M, Ohsaki Y, Ayukawa Y, Kido MA, Inoue T, Shimono M, Koyano K, Tanaka T. Ultrastructural and immunoelectron microscopic studies of the peri-implant epithelium-implant (Ti-6Al-4V) interface of rat maxilla. *J. Periodontol.* 2000;71:961-973.

[656] Uchida M, Oyane A, Kim HM, Kokubo T, Ito A. Biomimetic coating of laminin-apatite composite on titanium metal and its excellent cell-adhesive properties. *Adv. Mater.* 2004;16:1071-1074.

[657] Oyane A, Uchida M, Ito A. Laminin-apatite composite coating to enhance cell adhesion to ethylene-vinyl alcohol copolymer. *J. Biomed. Mater. Res. A* 2005;72A:168-174.

[658] Oyane A, Uchida M, Onuma K, Ito A. Spontaneous growth of a laminin-apatite nano-composite in a metastable calcium phosphate solution. *Biomaterials* 2006;27:167-175.

[659] Liu WB, Qu SX, Shen R, Jiang CX, Li XH, Feng B, Weng J. Influence of pH values on preparation of hydroxyapatite/gelatin composites. *J. Mater. Sci.* 2006;41:1851-1853.

[660] Yaylaoglu MB, Korkusuz P, Ors U, Korkusuz F, Hasirci V. Development of a calcium phosphate-gelatin composite as a bone substitute and its use in drug release. *Biomaterials* 1999;20:711-719.

[661] Kim HW, Song JH, Kim HE. Nanofiber generation of gelatin-hydroxyapatite biomimetics for guided tissue regeneration. *Adv. Funct. Mater.* 2005;15:1988-1994.

[662] Sivakumar M, Panduranga Rao K. Preparation, characterization and *in vitro* release of gentamicin from coralline hydroxyapatite-gelatin composite microspheres. *Biomaterials* 2002;23:3175-3181.

[663] Kim HW, Knowles JC, Kim HE. Porous scaffolds of gelatin-hydroxyapatite nanocomposites obtained by biomimetic approach: characterization and antibiotic drug release. *J. Biomed. Mater. Res. B Appl. Biomater.* 2005;74B:686-698.

[664] Yin YJ, Zhao F, Song XF, Yao KD, Lu WW, Leong JC. Preparation and characterization of hydroxyapatite/chitosan-gelatin network composite. *J. Appl. Polym. Sci.* 2000;77:2929-2938.

[665] Kim HW, Yoon BH, Kim HE. Microsphere of apatite-gelatin nanocomposite as bone regenerative filler. *J. Mater. Sci. Mater. Med.* 2005;16:1105-1109.

[666] Hillig WB, Choi Y, Murtha S, Natravali N, Ajayan P. An open-pored gelatin/hydroxyapatite composite as a potential bone substitute. *J. Mater. Sci. Mater. Med.* 2008;19:11-17.

[667] Chang MC, Douglas WH, Tanaka J. Organic-inorganic interaction and the growth mechanism of hydroxyapatite crystals in gelatin matrices between 37 and 80 °C. *J. Mater. Sci. Mater. Med.* 2006;17:387-396.

[668] Chang MC, Douglas WH. Cross-linkage of hydroxyapatite/gelatin nanocomposite using imide-based zero-length cross-linker. *J. Mater. Sci. Mater. Med.* 2007;18:2045-2051.

[669] Teng S, Chen L, Guo Y, Shi J. Formation of nano-hydroxyapatite in gelatin droplets and the resulting porous composite microspheres. *J. Inorg. Biochem.* 2007;101:686-691.

[670] Zhao F, Grayson WL, Ma T, Bunnell B, Lu WW. Effects of hydroxyapatite in 3-D chitosan-gelatin polymer network on human

mesenchymal stem cell construct development. *Biomaterials* 2006;27:1859-1867.
[671] Lin HR, Yeh YJ. Porous alginate/hydroxyapatite composite scaffolds for bone tissue engineering: preparation, characterization and *in vitro* studies. *J. Biomed. Mater. Res. B Appl. Biomater.* 2004;71B:52-65.
[672] Gelinsky M, Eckert M, Despang F. Biphasic, but mololithic scaffolds for the therapy of osteochondral defects. *Int. J. Mater. Res.* (formerly Z. Metallkd.) 2007;98:749-755.
[673] Yamaguchi I, Tokuchi K, Fukuzaki H, Koyama Y, Takakuda K, Monma H, Tanaka J. Preparation and microstructure analysis of chitosan/hydroxyapatite nanocomposites. *J. Biomed. Mater. Res.* 2001;55:20-27.
[674] Zhao F, Yin YJ, Lu WW, Leong JC, Zhang WJ, Zhang JY, Zhang MF, Yao KD. Preparation and histological evaluation of biomimetic three-dimensional hydroxyapatite/chitosan-gelatin network composite scaffolds. *Biomaterials* 2002;23:3227-3234.
[675] Shen X, Tong H, Jiang T, Zhu Z, Wan P, Hu J. Homogeneous chitosan/carbonate apatite/citric acid nanocomposites prepared through a novel in situ precipitation method. *Compos. Sci. Technol.* 2007;67:2238-2245.
[676] Murugan R, Ramakrishna S. Bioresorbable composite bone paste using polysaccharide based nano-hydroxyapatite. *Biomaterials* 2004;25:3829-3835.
[677] Yoshida A, Miyazaki T, Ishida E, Ashizuka M. Preparation of bioactive chitosan-hydroxyapatite nanocomposites for bone repair through mechanochemical reaction. *Mater. Trans.* 2004;45:994-998.
[678] Zhang Y, Ni M, Zhang MQ, Ratner B. Calcium phosphate – chitosan composite scaffolds for bone tissue engineering. *Tissue Eng.* 2003;9:337-345.
[679] Zhang Y, Zhang MQ. Microstructural and mechanical characterization of chitosan scaffolds reinforced by calcium phosphates. *J. Non-Cryst. Solids* 2001;282:159-164.
[680] Zhang Y, Zhang MQ. Calcium phosphate/chitosan composite scaffolds for controlled *in vitro* antibiotic drug release. *J. Biomed. Mater. Res.* 2002;62:378-386.
[681] Tachaboonyakiat W, Ogomi D, Serizawa T, Akashi M. Evaluation of cell adhesion and proliferation on a novel tissue engineering scaffold containing chitosan and hydroxyapatite. *J. Bioactive and Compatible Polymers* 2006;21:579-589.

[682] Sreedhar B, Aparna Y, Sairam M, Hebalkar N. Preparation and characterization of HAP/carboxymethyl chitosan nanocomposites. *J. Appl. Polym. Sci.* 2007;105:928-934.
[683] Rusu VM, Ng CH, Wilke M, Tiersch B, Fratzl P, Peter MG. Size-controlled hydroxyapatite nanoparticles as self-organized organic-inorganic composite materials. *Biomaterials* 2005;26:5414-5426.
[684] Pinheiro AG, Pereira FFM, Santos MRP, Freire FNA, Góes JC, Sombra ASB. Chitosan-hydroxyapatite-BIT composite films: preparation and characterization. *Polym. Compos.* 2007;28:582-587.
[685] Wan ACA, Khor E, Hastings GW. Preparation of a chitin-apatite composite by *in situ* precipitation onto porous chitin scaffolds. *J. Biomed. Mater. Res.* 1998;41:541-548.
[686] Wan ACA, Khor E, Hastings GW. Hydroxyapatite modified chitin as potential hard tissue substitute material. *J. Biomed. Mater. Res.* 1997;38:235-241.
[687] Geçer A, Yldz N, Erol M, Çalml A. Synthesis of chitin calcium phosphate composite in different growth media. Polymer Composites 2008;29:84-91.
[688] Dong H, Ye JD, Wang XP, Yang JJ. Preparation of calcium phosphate cement tissue engineering scaffold reinforced with chitin fiber. *J. Inorg. Mater.* 2007;22:1007-1010.
[689] Wang X, Ma J, Feng Q, Cui FZ. Skeletal repair in rabbits with calcium phosphate cements incorporated phosphorylated chitin. *Biomaterials* 2002;23:4591-4600.
[690] Wen HB, de Wijn JR, van Blitterswijk CA, de Groot K. Incorporation of bovine serum albumin in calcium phosphate coating on titanium. *J. Biomed. Mater. Res.* 1999;46:245-252.
[691] Liu TY, Chen SY, Liu DM, Liou SC. On the study of BSA-loaded calcium-deficient hydroxyapatite nano-carriers for controlled drug delivery. *J. Controlled Release* 2005;107:112-121.
[692] Liu Y, Hunziker E, Randall N, de Groot K, Layrolle P. Proteins incorporated into biomimetically prepared calcium phosphate coatings modulate their mechanical strength and dissolution rate. *Biomaterials* 2003;24:65-70.
[693] Dorozhkin SV, Dorozhkina EI. The influence of bovine serum albumin on the crystallization of calcium phosphates from a revised simulated body fluid. Colloids and Surfaces A: Physiochem. *Eng. Aspects* 2003;215:191-199.

[694] Fu HH, Hu YH, McNelis T, Hollinger JO. A calcium phosphate-based gene delivery system. *J. Biomed. Mater. Res. A* 2005;74A:40-48.
[695] Bisht S, Bhakta G, Mitra S, Maitra A. pDNA loaded calcium phosphate nanoparticles: highly efficient non-viral vector for gene delivery. *Int. J. Pharm.* 2005;288:157-168.
[696] Kakizawa Y, Miyata K, Furukawa S, Kataoka K. Size-controlled formation of a calcium phosphate-based organic-inorganic hybrid vector for gene delivery using poly(ethylene glycol)-block-poly(aspartic acid). *Adv. Mater.* 2004;16:699-702.
[697] Singh R, Saxena A, Mozumdar S. Calcium phosphate – DNA nanocomposites: morphological studies and their bile duct infusion for liver-directed gene therapy. *Int. J. Appl. Ceram. Technol.* 2008;5:1-10.
[698] Taguchi T, Kishida A, Akashi M. Hydroxyapatite formation on/in poly(vinyl alcohol) hydrogel matrices using a novel alternate soaking process. *Chem. Lett.* 1998;8:711-712.
[699] Tachaboonyakiat W, Serizawa T, Akashi M. *Hydroxyapatite formation on/in iodegradable chitosan hydrogels by an alternate soaking process. Polym. J.* 2001;33:177-181.
[700] Schnepp ZAC, Gonzalez-McQuire R, Mann S. Hybrid biocomposites based on calcium phosphate mineralization of self-assembled supramolecular hydrogels. *Adv. Mater.* 2006;18:1869-1872.
[701] Bigi A, Boanini E, Gazzano M, Kojdecki MA, Rubini K. Microstructural investigation of hydroxyapatite-polyelectrolyte composites. *J. Mater. Chem.* 2004;14:274-279.
[702] Bigi A, Boanini E, Gazzano M, Rubini K, Torricelli P. Nanocrystalline hydroxyapatite-polyaspartate composites. *Biomed. Mater. Eng.* 2004;14:573-579.
[703] Boanini E, Fini M, Gazzano M, Bigi A. Hydroxyapatite nanocrystals modified with acidic amino acids. *Eur. J. Inorg. Chem.* 2006;4821-4826.
[704] Boanini E, Torricelli P, Gazzano M, Giardino R, Bigi A. Nanocomposites of hydroxyapatite with aspartic acid and glutamic acid and their interaction with osteoblast-like cells. *Biomaterials* 2006;27:4428-4433.
[705] Sánchez-Salcedo S, Nieto A, Vallet-Regi M. Hydroxyapatite/β-tricalcium phosphate/agarose macroporous scaffolds for bone tissue engineering. *Chem. Engin. J.* 2005;137:62-71.
[706] Román J, Cabañas MV, Peña J, Doadrio JC, Vallet-Regi M. An optimized β-tricalcium phosphate and agarose scaffold fabrication technique. *J. Biomed. Mater. Res. A* 2008;84A:99-107.

[707] Abiraman S, Varma H, Umashankar P, John A. Fibrin sealant as an osteoinductive protein in a mouse model. *Biomaterials* 2002;23:3023-3031.
[708] Bagot d'Arc M, Daculsi G. Micro macroporous biphasic ceramics and fibrin sealant as a mouldable material for bone reconstruction in chronic otitis media surgery: a 15 years experience. *J. Mater. Sci. Mater. Med.* 2003;14:229-233.
[709] Bonucci E, Marini E, Valdinucci F, Fortunato G. Osteogenic response to hydroxyapatite-fibrin implants in maxillofacial bone defects. *Eur. J. Oral Sci.* 1997;105:557-561.
[710] Fortunato G, Marini E, Valdinucci F, Bonucci E. Long-term results of hydroxyapatite-fibrin sealant implantation in plastic and reconstructive craniofacial surgery. *J. Cranio Maxillofac. Surg.* 1997;25:124-135.
[711] Jegoux F, Goyenvalle E, Bagot d'Arc M, Aguado E, Daculsi G. *In vivo* biological performance of composites combining micro-macroporous biphasic calcium phosphate granules and fibrin sealant. *Arch. Orthop. Trauma Surg.* 2005;125:153-159.
[712] Le Guehennec L, Goyenvalle E, Aguado E, Pilet P, Bagot d'Arc M, Daculsi G. MBCP® biphasic calcium phosphate granules and Tissucol® fibrin sealant in rabbit femoral defects: the effect of fibrin on bone ingrowth. *J. Mater. Sci. Mater. Med.* 2005;16:29-35.
[713] Wittkampf A. Fibrin sealant as sealant for hydroxyapatite granules. *J. Cranio Maxillofac. Surg.* 1989;17:179-181.
[714] Le Nihouannen D, Guehennec LL, Rouillon T, Pilet P, Bilban M, Layrolle P, Daculsi G. Micro-architecture of calcium phosphate granules and fibrin glue composites for bone tissue engineering. *Biomaterials* 2006;27:2716-2722.
[715] Le Nihouannen D, Saffarzadeh A, Aguado E, Goyenvalle E, Gauthier O, Moreau F, Pilet P, Spaethe R, Daculsi G, Layrolle P. Osteogenic properties of calcium phosphate ceramics and fibrin glue based composites. *J. Mater. Sci. Mater. Med.* 2007;18:225-235.
[716] Le Guehennec L, Goyenvalle E, Aguado E, Pilet P, Spaethe R, Daculsi G. Influence of calcium chloride and aprotinin in the *in vivo* biological performance of a composite combining biphasic calcium phosphate granules and fibrin sealant. *J. Mater. Sci. Mater. Med.* 2007;18:1489-1495.
[717] Le Nihouannen D, Goyenvalle E, Aguado E, Pilet P, Bilban M, Daculsi G, Layrolle P. Hybrid composites of calcium phosphate granules, fibrin

glue, and bone marrow for skeletal repair. *J. Biomed. Mater. Res. A* 2007;81A:399-408.
[718] Le Nihouannen D, Saffarzadeh A, Gauthier O, Moreau F, Pilet P, Spaethe R, Layrolle P, Daculsi G. Bone tissue formation in sheep muscles induced by a biphasic calcium phosphate ceramic and fibrin glue composite. *J. Mater. Sci. Mater. Med.* 2008;19:667-675.
[719] Yoh R, Matsumoto T, Sasaki JI, Sohmura T. Biomimetic fabrication of fibrin/apatite composite material. *J. Biomed. Mater. Res. A* 2008;87A:222-228.
[720] Boanini E, Torricelli P, Gazzano M, Giardino R, Bigi A. Alendronate-hydroxyapatite nanocomposites and their interaction with osteoclasts and osteoblast-like cells. *Biomaterials* 2008;29:790-796.
[721] Li L, Wei KM, Lin F, Kong XD, Yao JM. Effect of silicon on the formation of silk fibroin/calcium phosphate composite. *J. Mater. Sci. Mater. Med.* 2008;19:577-582.
[722] Wang L, Nemoto R, Senna M. Effects of alkali pretreatment of silk fibroin on microstructure and properties of hydroxyapatite-silk fibroin nanocomposite. *J. Mater. Sci. Mater. Med.* 2004;15:261-265.
[723] Wang L, Nemoto R, Senna M. Microstructure and chemical states of hydroxyapatite/silk fibroin nanocomposites synthesized via a wet-mechanochemical route. *J. Nanoparticle Res.* 2002;4:535-540.
[724] Wang L, Nemoto R, Senna M. Three-dimensional porous network structure developed in hydroxyapatite-based nanocomposites containing enzyme pretreated silk fibroin. *J. Nanoparticle Res.* 2004;6:91-98.
[725] Nemoto R, Wang L, Ikoma T, Tanaka J, Senna M. Preferential alignment of hydroxyapatite crystallites in nanocomposites with chemically disintegrated silk fibroin. *J. Nanoparticle Res.* 2004;6:259-265.
[726] Wang L, Li CZ, Senna M. High-affinity integration of hydroxyapatite nanoparticles with chemically modified silk fibroin. *J. Nanoparticle Res.* 2007;9:919-929.
[727] Wang L, Li CZ. Preparation and physicochemical properties of a novel hydroxyapatite/chitosan-silk fibroin composite. *Carbohydr. Polym.* 2007;68:740-745.
[728] Sogo Y, Ito A, Matsuno T, Oyane A, Tamazawa G, Satoh T, Yamazaki A, Uchimura E, Ohno T. Fibronectin-calcium phosphate composite layer on hydroxyapatite to enhance adhesion, cell spread and osteogenic differentiation of human mesenchymal stem cells *in vitro*. *Biomed. Mater.* 2007;2:116-123.

[729] Cross KJ, Huq NL, Palamara JE, Perich JW, Reynolds EC. Physicochemical characterization of casein phosphopeptide-amorphous calcium phosphate nanocomplexes. *J. Biol. Chem.* 2005;280:15362-15369.

[730] Shchukin DG, Sukhorukov GB, Möhwald H. Biomimetic fabrication of nanoengineered hydroxyapatite/polyelectrolyte composite shell. *Chem. Mater.* 2003;15:3947-3950.

[731] Weiss P, Gauthier O, Bouler JM, Grimandi G, Daculsi G. Injectable bone substitute using a hydrophilic polymer. *Bone* 1999;25(2 Suppl):67S-70S.

[732] Daculsi G, Weiss P, Bouler JM, Gauthier O, Millot F, Aguado E. Biphasic calcium phosphate/hydrosoluble polymer composites: a new concept for bone and dental substitution biomaterials. *Bone* 1999;25(2 Suppl):59S-61S.

[733] Turczyn R, Weiss P, Lapkowski M, Daculsi G. *In situ* self-hardening bioactive composite for bone and dental surgery. *J. Biomaterials Sci. Polymer Ed.* 2000;11:217-223.

[734] Bennett S, Connolly K, Lee DR, Jiang Y, Buck D, Hollinger JO, Gruskin EA. Initial biocompatibility studies of a novel degradable polymeric bone substitute that hardens *in situ*. *Bone* 1996;19:101S-107S.

[735] Daculsi G, Rohanizadeh R, Weiss P, Bouler JM. Crystal polymer interaction with new injectable bone substitute; SEM and HrTEM study. *J. Biomed. Mater. Res.* 2000;50:1-7.

[736] Grimande G, Weiss P, Millot F, Daculsi G. *In vitro* evaluation of a new injectable calcium phosphate material. *J. Biomed. Mater. Res.* 1998;39:660-666.

[737] Weiss P, Lapkowski M, LeGeros RZ, Bouler JM, Jean A, Daculsi G. FTIR spectroscopic study of an organic/mineral composite for bone and dental substitute materials. *J. Mater. Sci. Mater. Med.* 1997;8:621-629.

[738] Weiss P, Bohic S, Lapkowski M, Daculsi G. Application of FTIR microspectroscopy to the study of an injectable composite for bone and dental surgery. *J. Biomed. Mater. Res.* 1998;41:167-170.

[739] Schmitt M, Weiss P, Bourges X, del Valle GA, Daculsi G. Crystallization at the polymer/calcium-phosphate interface in a sterilized injectable bone substitute IBS. *Biomaterials* 2002;23:2789-2794.

[740] Gauthier O, Müller R, von Stechow D, Lamy B, Weiss P, Bouler JM, Aguado E, Daculsi G. *In vivo* bone regeneration with injectable calcium phosphate biomaterial: a three-dimensional micro-computed

tomographic, biomechanical and SEM study. *Biomaterials* 2005;26:5444-5453.

[741] Weiss P, Layrolle P, Clergeau LP, Enckel B, Pilet P, Amouriq Y, Daculsi G, Giumelli B. The safety and efficacy of an injectable bone substitute in dental sockets demonstrated in a human clinical trial. *Biomaterials* 2007;28:3295-3305.

[742] Trojani C, Boukhechba F, Scimeca JC, Vandenbos F, Michiels JF, Daculsi G, Boileau P, Weiss P, Carle GF, Rochet N. Ectopic bone formation using an injectable biphasic calcium phosphate/Si-HPMC hydrogel composite loaded with undifferentiated bone marrow stromal cells. *Biomaterials* 2006;27:3256-3264.

[743] Iooss P, Le Ray AM, Grimandi G, Daculsi G, Merle C. A new injectable bone substitute combining poly(ε-caprolactone) microparticles with biphasic calcium phosphate granules. *Biomaterials* 2001;22:2785-2794.

[744] Evis Z, Ergun C, Doremus RH. Hydroxylapatite-zirconia composites: thermal stability of phases and sinterability as related to the CaO-ZrO_2 phase diagram. *J. Mater. Sci.* 2005;40:1127-1134.

[745] Rao RR, Kannan TS. Synthesis and sintering of hydroxyapatite-zirconia composites. *Mater. Sci. Eng. C* 2002;20:187-193.

[746] Mansur C, Pope M, Pascucci MR, Shivkumar S. Zirconia-calcium phosphate composites for bone replacement. *Ceram. Int.* 1998;24:77-79.

[747] Kim HW, Kim HE, Salih V, Knowles JC. Dissolution control and cellular responses of calcium phosphate coatings on zirconia porous scaffold. *J. Biomed. Mater. Res. A* 2004;68A:522-530.

[748] Milella E, Cosentino F, Licciulli A, Massaro C. Preparation and characterisation of titania/hydroxyapatite composite coatings obtained by sol-gel process. *Biomaterials* 2001;22:1425-1431.

[749] Goller G, Demirkiran H, Oktar FN, Demirkesen E. Processing and characterization of Bioglass reinforced hydroxyapatite composites. *Ceram. Int.* 2003;29:721-724.

[750] Tancred DC, Carr AJ, McCormack BA. The sintering and mechanical behav*iour of hydroxyapatite with bioglass additions.* J. Mater. Sci. Mater. Med. 2001;12:81-93.

[751] Lopes MA, Silva RF, Monteiro FJ, Santos JD. Microstructural dependence of Young's moduli of P_2O_5 glass reinforced hydroxyapatite for biomedical applications. *Biomaterials* 2000;21:749-754.

[752] Juang HY, Hon MH. Fabrication and mechanical properties of hydroxyapatite-alumina composites. *Mater. Sci. Eng. C* 1994;2:77-81.

[753] Li J, Forbreg S, Hermansson L. Evaluation of the mechanical properties of hot isotatically pressed titania and titania-calcium phosphate composites. *Biomaterials* 1991;12:438-440.

[754] Noma T, Shoji N, Wada S, Suzuki T. Preparation of spherical Al_2O_3 particle dispersed hydroxyapatite ceramics *J. Ceram. Soc. Japa*n 1993;101:923-927.

[755] Gautier S, Champion E, Bernache-Assollant D. Toughening characterization in alumina platelet-hydroxyapatite matrix composites. *J. Mater. Sci. Mater. Med.* 1999;10:533-540.

[756] Fang Y, Roy DM, Cheng J, Roy R, Agrawal DK. Microwave sintering of hydroxyapatite-based composites. *Ceram. Trans.* 1993;36:397-407.

[757] Park K, Vasilosa T. Microstructure and mechanical properties of silicon carbide whisker/calcium phosphate composites produced by hot pressing. *Mater. Lett.* 1997;32:229-233.

[758] de With G, Corbijn AT. Metal fibre reinforced hydroxyapatite ceramics. *J. Mater. Sci.* 1989;24:3411-3415.

[759] Ruys AJ, Simpson SA, Sorrell CC. Thixotropic casting of fibre-reinforced ceramic matrix composites. *J. Mater. Sci. Lett.* 1994;13:1323-1325.

[760] Miao X, Ruys AJ, Milthorpe BK. Hydroxyapatite-316L fibre composites prepared by vibration assisted slip casting. *J. Mater. Sci.* 2001;36:3323-3332.

[761] Li J, Liao H, Hermansson L. Sintering of partially-stabilized zirconia and partially-stabilized zirconia-hydroxyapatite composites by hot isostatic pressing and pressureless sintering. *Biomaterials* 1996;17:1787-1790.

[762] Takagi M, Mochida M, Uchida N, Saito K, Uematsu K. Filter cake forming and hot isostatic pressing for TZP-dispersed hydroxyapatite composite. *J. Mater. Sci. Mater. Med.* 1992;3:199-203.

[763] Silva VV, Domingues RZ. Hydroxyapatite-zirconia composites prepared by precipitation method. *J. Mater. Sci. Mater. Med.* 1997;8:907-910.

[764] Silva VV, Lameiras FS, Domingues RZ. Synthesis and characterization of calcia partially stabilized zirconia-hydroxyapatite powders prtepared by co-precipitation method. *Ceram. Int.* 2001;27:615-620.

[765] Rapacz-Kmita A, Slosarczyk A, Paszkiewicz Z, Paluch D. Evaluation of HAp-ZrO_2 composites and monophase HAp bioceramics. *In vitro* study. *J. Mater. Sci.* 2004;39:5865-5867.

[766] Rapacz-Kmita A, Slosarczyk A, Paszkiewicz Z. HAp ZrO_2 composite coatings prepared by plasma spraying for biomedical applications. *Ceram. Int.* 2005;31:567-571.
[767] Sung YM, Kim DH. Crystallization characteristics of yttria-stabilized zirconia/hydroxyapatite composite nanopowder. *J. Crystal Growth* 2003;254:411-417.
[768] Silva VV, Lameiras FS. Synthesis and characterization of composite powders of partially stabilized zirconia and hydroxyapatite. *Mater. Character.* 2000;45:51-59.
[769] Shen Z, Adolfsson E, Nygren M, Gao L, Kawaoka H, Niihara K. Dense hydroxyapatite-zirconia ceramic composites with high strength for biological applications. *Adv. Mater.* 2001;13:214-216.
[770] Adolfsson E, Alberiushenning P, Hermansson L. Phase-analysis and thermal-stability of hot isostatically pressed zirconia-hydroxyapatite composites. *J. Am. Ceram. Soc.* 2000;83:2798-2802.
[771] Kim HW, Noh YJ, Koh YH, Kim HE, Kim HM. Effect of CaF_2 on densification and properties of hydroxyapatite-zirconia composites for biomedical applications. *Biomaterials* 2002;23:4113-4121.
[772] Li W, Gao L. Fabrication of HAp-ZrO_2 (3Y) nano-composite by SPS. *Biomaterials* 2003;24:937-940.
[773] Kim HW, Knowles JC, Li LH, Kim HE. Mechanical performance and osteoblast-like cell responses of fluorine-substituted hydroxyapatite and zirconia dense composite. *J. Biomed. Mater. Res. A* 2005;72A:258-268.
[774] Xiao XF, Liu RF, Zheng YZ. Hydrothermal-electrochemical codeposited hydoxyapatite/yttria-stabilized zirconia composite coating. *J. Mater. Sci.* 2006;41:3417-3424.
[775] Kumar BR, Prakash KH, Cheang P, Khor KA. Microstructure and mechanical properties of spark plasma sintered zirconia-hydroxyapatite nano-composite powders. *Acta Materialia* 2005;53:2327-2335.
[776] Ahn ES, Gleason NJ, Nakahira A, Ying JY. Nanostructure processing of hydroxyapatite-based bioceramics. *Nano Lett.* 2001;1:149-153.
[777] Silva VV, Lameiras FS, Domingues RZ. Microstructural and mechanical study of zirconia-hydroxyapatite (ZH) composite ceramics for biomedical applications. *Compos. Sci. Technol.* 2001;61:301-310.
[778] Khor KA, Fu L, Lim JP, Cheang P. The effects of ZrO_2 on the phase compositions of plasma sprayed HA/YSZ composite coatings. *Mater. Sci. Eng. A* 2001;316:160-166.

[779] Fu L, Khor KA, Lim JP. Effects of yttria-stabilized zirconia on plasma-sprayed hydroxyapatite/yttria-stabilized zirconia composite coatings. *J. Am. Ceram. Soc.* 2002;85:800-806.
[780] Chou BY, Chang E, Yao SY, Chen JM. Phase transformation during plasma spraying of hydroxyapatite-10-wt%-zirconia composite coating. *J. Am. Ceram. Soc.* 2002;85:661-669.
[781] Wang Q, Ge S, Zhang D. Nano-mechanical properties and biotribological behaviors of nanosized HA/partially-stabilized zirconia composites. *Wear* 2005;259:952-957.
[782] Murugan R, Ramakrishna S.Effect of zirconia on the formation of calcium phosphate bioceramics under microwave irradiation. *Mater. Lett.* 2003;58:230-234.
[783] Fu L, Khor KA, Lim JP. Processing, microstructure and mechanical properties of yttria stabilized zirconia reinforced hydroxyapatite coatings. *Mater. Sci. Eng. A* 2000;276:46-51.
[784] Nagarajan VS, Rao KJ. Structural, mechanical and biocompatibility studies of hydroxyapatite-derived composites toughened by zirconia addition. *J. Mat. Chem.* 1993;3:43-51.
[785] Tamari N, Kondo I, Mouri M, Kinoshita M. Effect of calcium fluoride addition on densification and mechanical properties of hydroxyapatite-zirconia composite ceramics. *J. Ceram. Soc. Japan* 1988;96:1200-1202.
[786] Evis Z, Doremus RH. Hot-pressed hydroxylapatite/monoclinic zirconia composites with improved mechanical properties. *J. Mater. Sci.* 2007;42:2426-2431.
[787] Evis Z, Doremus RH. Effect of MgF_2 on hot pressed hydroxylapatite and monoclinic zirconia composites. *J. Mater. Sci.* 2007;42:3739-3744.
[788] Ahn ES, Gleason NJ, Ying JY. The effect of zirconia reinforcing agents on the microstructure and mechanical properties of hydroxyapatite-based nanocomposites. *J. Am. Ceram. Soc.* 2005;88:3374-3379.
[789] Erkmen ZE, Genç Y, Oktar FN. Microstructural and mechanical properties of hydroxyapatite-zirconia composites. *J. Am. Ceram. Soc.* 2007;90:2885-2892.
[790] Rapacz-Kmita A, Slosarczyk A, Paszkiewicz Z. Mechanical properties of HAp-ZrO_2 composites. *J. Eur. Ceram. Soc.* 2006;26:1481-1488.
[791] Sung YM, Shin YK, Ryu JJ. Preparation of hydroxyapatite/zirconia bioceramic nanocomposites for orthopaedic and dental prosthesis applications. *Nanotechnology* 2007;18:065602 (6 pp.).

[792] Quan R, Yang D, Wu X, Wang H, Miao X. Li W. *In vitro* and *in vivo* biocompatibility of graded hydroxyapatite-zirconia composite bioceramic. *J. Mater. Sci. Mater. Med.* 2008:19:183-187.
[793] Khalil KA, Kim SW, Kim HY. Consolidation and mechanical properties of nanostructured hydroxyapatite-(ZrO_2 + 3 mol% Y_2O_3) bioceramics by high-frequency induction heat sintering. *Mater. Sci. Eng. A* 2007;456:368-372.
[794] Kong YM, Kim S, Kim HE, Lee IS. Reinforcement of hydroxyapatite bioceramic by addition of ZrO_2 coated with Al_2O_3. *J. Am. Ceram. Soc.* 1999;82:2963-2968.
[795] Choi JW, Kong YM, Kim HE, Lee IS. Reinforcement of hydroxyapatite bioceramic by addition of Ni_3Al and Al_2O_3. *J. Am. Ceram. Soc.* 1998;81:1743-1748.
[796] Adolfsson E, Hermansson L. Phase stability aspects of various apatite-aluminium oxide composites. *J. Mater. Sci.* 2000;35:5719-5723.
[797] Li J, Fartash B, Hermansson L. Hydroxyapatite-alumina composites and bone-bonding. *Biomaterials* 1995;16:417-422.
[798] Kim S, Kong YM, Kim HE, Lee IS. Effect of calcinations of starting powder on mechanical properties of hydroxyapatite-alumina bioceramic composite. *J. Mater. Sci. Mater. Med.* 2002;13:307-310.
[799] Chiba A, Kimura S, Raghukandan K, Morizono Y. Effect of alumina addition on hydroxyapatite biocomposites fabricated by underwater-shock compaction. *Mater. Sci. Eng. A* 2003;350:179-183.
[800] Pang YX, Bao X, Weng L. Preparation of tricalcium phosphate/alumina composite nanoparticles and self-reinforcing composites by simultaneous precipitation. *J. Mater. Sci.* 2004;39:6311-6323.
[801] Jun YK, Kim WH, Kweon OK, Hong SH. The fabrication and biochemical evaluation of alumina reinforced calcium phosphate porous implants. *Biomaterials* 2003;24:3731-3739.
[802] Epure LM, Dimitrievska S, Merhi Y, Yahia LH. The effect of varying Al_2O_3 percentage in hydroxyapatite/Al_2O_3 composite materials: morphological, chemical and cytotoxic evaluation. *J. Biomed. Mater. Res. A* 2007;83A:1009-1023.
[803] Evis Z, Doremus RH. A study of phase stability and mechanical properties of hydroxylapatite-nanosize α-alumina composites. *Mater. Sci. Eng. C* 2007;27:421-425.
[804] Lu YP, Li MS, Li ST, Wang ZG, Zhu RF. Plasma-sprayed hydroxyapatite+titania composite bond coat for hydroxyapatite coating on titanium substrate. *Biomaterials* 2004;25:4393-4403.

[805] Li H., Khor KA, Cheang P. Impact formation and microstructure characterization of thermal sprayed hydroxyapatite/titania composite coatings. *Biomaterials* 2003;24:949-957.

[806] Zheng XB, Ding CX. Characterization of plasma-sprayed hydroxyapatite/TiO$_2$ composite coatings. *J. Therm. Spray Technol.* 2000;9:520-525.

[807] Lee SH, Kim HW, Lee EJ, Li LH, Kim HE. Hydroxyapatite-TiO$_2$ hybrid coating on Ti implants. *J. Biomater. Appl.* 2006;20:195-208.

[808] Lin C, Yen S. Characterization and bond strength of electrolytic HA/TiO$_2$ double layers for orthopedic applications. *J. Mater. Sci. Mater. Med.* 2004;15:1237-1246.

[809] Balamurugan A, Balossier G, Kannan S, Michel J, Rajeswari S. *In vitro* biological, chemical and electrochemical evaluation of titania reinforced hydroxyapatite sol-gel coatings on surgical grade 316L SS. *Mater. Sci. Eng. C* 2007;27:162-171.

[810] Gaona M, Limab RS, Marple BR. Nanostructured titania/hydroxyapatite composite coatings deposited by high velocity oxy-fuel (HVOF) spraying. *Mater. Sci. Eng. A* 2007;458:141-149.

[811] Boyd AR, Duffy H, McCann R, Meenan BJ. Sputter deposition of calcium phosphate/titanium dioxide hybrid thin films. *Mater. Sci. Eng. C* 2008;28:228-236.

[812] Fidancevska E, Ruseska G, Bossert J, Linc YM, Boccaccini AR. Fabrication and characterization of porous bioceramic composites based on hydroxyapatite and titania. *Mater. Chem. Phys.* 2007;103:95-100.

[813] Harle J, Kim HW, Mordan N, Knowles JC, Salih V. Initial responses of human osteoblasts to sol-gel modified titanium with hydroxyapatite and titania composition. *Acta Biomater.* 2006;2:547-556.

[814] Kim HW, Kim HE, Salih V, Knowles JC. Hydroxyapatite and titania sol-gel composite coatings on titanium for hard tissue implants; mechanical and *in vitro* biological performance. *J. Biomed. Mater. Res. B Appl. Biomater.* 2005;72B:1-8.

[815] Pushpakanth S, Srinivasan B, Sreedhar B, Sastry TP. An *in situ* approach to prepare nanorods of titania-hydroxyapatite (TiO$_2$-HAp) nanocomposite by microwave hydrothermal technique. *Mater. Chem. Phys.* 2008;107:492-498.

[816] Sato M, Aslani A, Sambito MA, Kalkhoran NM, Slamovich EB, Webster TJ. Nanocrystalline hydroxyapatite/titania coatings on titanium improves osteoblast adhesion. *J. Biomed. Mater. Res.* 2008;84A:265-272.

[817] Ramires PA, Romito A, Cosentino F, Milella E. The influence of titania/hydroxyapatite composite coatings on *in vitro* osteoblasts behaviour. *Biomaterials* 2001;22:1467-1474.

[818] Sun R, Li M, Lu Y, An X. Effect of titanium and titania on chemical characteristics of hydroxyapatite plasma-sprayed into water. *Mater. Sci. Eng. C* 2006;26:28-33.

[819] Lee BT, Lee CW, Gain AK, Song HY. Microstructures and material properties of fibrous Ap/Al_2O_3-ZrO_2 composites fabricated by multipass extrusion process. *J. Eur. Ceram. Soc.* 2007;27:157-163.

[820] Kong YM, Bae CJ, Lee SH, Kim HW, Kim HE. Improvement in biocompatibility of ZrO_2-Al_2O_3 nano-composite by addition of HA. *Biomaterials* 2005;26:509-517.

[821] Oktar FN, Agathopoulos S, Ozyegin LS, Gunduz O, Demirkol N, Bozkurt Y, Salman S. Mechanical properties of bovine hydroxyapatite (BHA) composites doped with SiO_2, MgO, Al_2O_3 and ZrO_2. *J. Mater. Sci. Mater. Med.* 2007;18:2137-2143.

[822] Gunduz O, Erkan EM, Daglilar S, Salman S, Agathopoulos S, Oktar FN. Composites of bovine hydroxyapatite (BHA) and ZnO. *J. Mater. Sci.* 2008;43:2536-2540.

[823] Li XW, Yasuda HY, Umakoshi Y. Bioactive ceramic composites sintered from hydroxyapatite and silica at 1200°C: preparation, microstructures and *in vitro* bone-like layer growth. *J. Mater. Sci. Mater. Med.* 2006;17:573-581.

[824] Ragel CV, Vallet-Regi M, Rodríguez-Lorenzo LM. Preparation and *in vitro* bioactivity of hydroxyapatite/sol-gel glass biphasic material. *Biomaterials* 2002;23:1865-1872.

[825] Padilla S, Sánchez-Salcedo S, Vallet-Regi M. Bioactive and biocompatible pieces of HA/sol-gel glass mixtures obtained by the gel-casting method. *J. Biomed. Mater. Res. A* 2005;75A:63-72.

[826] Lopes MA, Monterio FJ, Santos JD. Glass-reinforced hydroxyapatite composites: fracture toughness and hardness dependence on microstructural characteristics. *Biomaterials* 1999;20:2085-2090.

[827] Fu Q, Zhou N, Huang W, Wang D, Zhang L, Li H. Preparation and characterization of a novel bioactive bone cement: glass based nanoscale hydroxyapatite bone cement. *J. Mater. Sci. Mater. Med.* 2004;15:1333-1338.

[828] Fu Q, Zhou N, Huang W, Wang D, Zhang L, Li H. Effects of nano HAP on biological and structural properties of glass bone cement. *J. Biomed. Mater. Res. A* 2005;74A:156-163.

[829] Oktar FN, Goller G. Sintering effects on mechanical properties of glass-reinforced hydroxyapatite composites. *Ceram. Int.* 2002;28:617-621.
[830] Padilla S, Román J, Sánchez-Salcedo S, Vallet-Regi M. Hydroxyapatite/SiO_2-CaO-P_2O_5 glass materials: *in vitro* bioactivity and biocompatibility. *Acta Biomater.* 2006;2:331-342.
[831] Kokubo T, Shigematsu M, Nagashima Y, Tashiro M, Nakamura T, Yamamuro T, Higashi S. *Apatite- and wollastonite-containing glass ceramics for prosthetic applications.* Bulletin of the Institute for Chemical Research, 60. Kyoto University, 1982, pp. 260-268.
[832] Kitsugi T, Yamamuro T, Nakamura T, Higashi S, Kakutani Y, Hyakuna K, Ito S, Kokubo T, Takagi M, Shibuya T. Bone bonding behavior of three kinds of apatite containing glass ceramics. *J. Biomed. Mater. Res.* 1986;20:1295-1307.
[833] Kokubo T, Ito S, Shigematsu M, Sakka S, Yamamuro T. Fatigue and life-time of bioactive glass-ceramic A-W containing apatite and wollastonite. *J. Mater. Sci.* 1987;22:4067-4070.
[834] Kokubo T, Ito S, Shigematsu M, Sakka S, Yamamuro T. Mechanical properties of a new type of apatite-containing glass-ceramic for prosthetic application. *J. Mater. Sci.* 1985;20:2001-2004.
[835] Kokubo T. Bioactive glass ceramics: properties and applications. *Biomaterials* 1991;12:155-163.
[836] Kokubo T, Ito S, Huang ZT, Hayashi T, Sakka S, Kitsugi T, Yamamuro T. CaP-rich layer formed on high strength bioactive glass ceramic A-W. *J. Biomed. Mater. Res.* 1990;24:331-343.
[837] Nishio K, Neo M, Akiyama H, Okada Y, Kokubo T, Nakamura T. Effects of apatite and wollastonite containing glass-ceramic powder and two types of alumina powder in composites on osteoblastic differentiation of bone marrow cells. *J. Biomed. Mater. Res.* 2001;55:164-176.
[838] Zhang D, Chang J, Zeng Y. Fabrication of fibrous poly(butylene succinate)/wollastonite/apatite composite scaffolds by electrospinning and biomimetic process. *J. Mater. Sci. Mater. Med.* 2008;19:443-449.
[839] Chaki TK, Wang PE. Densification and strengthening of silver-reinforced hydroxyapatite-matrix composite prepared by sintering. *J. Mater. Sci. Mater. Med.* 1994;5:533-542.
[840] Zhang X, Gubbels GHM, Terpstra RA, Metselaar R. Toughening of calcium hydroxyapatite with silver particles. *J. Mater. Sci.* 1997;32:235-243.

[841] Chu C, P. Lin, Y. Dong, X. Xue, J. Zhu, Yin Z. Fabrication and characterization of hydroxyapatite reinforced with 20 vol % Ti particles for use as hard tissue replacement. *J. Mater. Sci. Mater. Med.* 2002;13:985-992.
[842] Shi W, Kamiya A, Zhu J, Watazu A. Properties of titanium biomaterial fabricated by sinter-bonding of titanium/hydroxyapatite composite surface-coated layer to pure bulk titanium. *Mater. Sci. Eng. A* 2002;337:104-109.
[843] Ning CQ, Zhou Y. *In vitro* bioactivity of a biocomposite fabricated from HA and Ti powders by powder metallurgy method. *Biomaterials* 2002;23:2909-2915.
[844] Karanjai M. Sundaresan R, Rao GVN, Rama Mohan TR, Kashyap BP. Development of titanium based biocomposite by powder metallurgy processing with in situ forming of Ca-P phases. *Mater. Sci. Eng. A* 2007;447:19-26.
[845] Karanjai M, Manoj Kumarb BV, Sundaresan R, Basu B, Rama Mohan TR, Kashyap BP. Fretting wear study on Ti-Ca-P biocomposite in dry and simulated body fluid. *Mater. Sci. Eng. A* 2008;475:299-307.
[846] Ning CQ, Zhou Y. On the microstructure of biocomposites sintered from Ti, HA and bioactive glass. *Biomaterials* 2004;25:3379-3387.
[847] Chu C, Xue X, Zhu J, Yin Z. Mechanical and biological properties of hydroxyapatite reinforced with 40 vol. % titanium particles for use as hard tissue replacement. *J. Mater. Sci. Mater. Med.* 2004;15:665-670.
[848] Smirnov VV, Egorov AA, Barinov SM, Shvorneva LI. Composite calcium phosphate bone cements reinforced by particulate titanium. *Doklady Chemistry* 2007;413:82-85.
[849] Chu C, Xue X, Zhu J, Yin Z. Fabrication and characterization of titanium-matrix composite with 20 vol% hydroxyapatite for use as heavy load-bearing hard tissue replacement. *J. Mater. Sci. Mater. Med.* 2006;17:245-251.
[850] Li J, Habibovic P, Yuan H, van den Doel M, Wilson CE, de Wijn JR, van Blitterswijk CA, de Groot K. Biological performance in goats of a porous titanium alloy-biphasic calcium phosphate composite. *Biomaterials* 2007;28:4209-4218.
[851] Pattanayak DK, Mathur V, Rao BT, Rama Mohan TR. Synthesis and characterization of titanium – calcium phosphate composites for bio applications. *Trends Biomater. Artif. Organs* 2003;17:8-12.
[852] Ding Y, Liu J, Wang H, Shen G, Yu R. A piezoelectric immunosensor for the detection of α-fetoprotein using an interface of

gold/hydroxyapatite hybrid nanomaterial. *Biomaterials* 2007;28:2147-2154.

[853] Damien CJ, Parsons JR, Benedict JJ, Weisman DS. Investigation of a hydroxyapatite and calcium sulfate composite supplemented with an osteoinductive factor. *J. Biomed. Mater. Res.* 1990;24:639-654.

[854] Rauschmann MA, Wichelhaus TA, Stirnal V, Dingeldein E, Zichner L, Schnettler R, Alt V. Nanocrystalline hydroxyapatite and calcium sulphate as biodegradable composite carrier material for local delivery of antibiotics in bone infections. *Biomaterials* 2005;26:2677-2684.

[855] Urban RM, Turner TM, Hall DJ, Inoue N, Gitelis S. Increased bone formation using calcium sulfate-calcium phosphate composite graft. *Clin. Orthop. Relat. Res.* 2007;459:110-117.

[856] Gittings JP, Bowena CR, Turner IG, Baxter F, Chaudhuri J. Characterisation of ferroelectric-calcium phosphate composites and ceramics. *J. Eur. Ceram. Soc.* 2007;27:4187-4190.

[857] Watanabe Y, IkomaT, Suetsugu Y, Yamada H, Tamura K, Komatsu Y, Tanaka J, Moriyoshi Y. The densification of zeolite/apatite composites using a pulse electric current sintering method: a long-term assurance material for the disposal of radioactive waste. *J. Eur. Ceram. Soc.* 2006;26:481-486.

[858] Agathopoulos S, Tulyaganov DU, Marques PAAP, Ferro MC, Fernandes MHV, Correia RN. The fluorapatite-anorthite system in biomedicine. *Biomaterials* 2003;24:1317-1331.

[859] Khor KA, Gu YW, Pan D, Cheang P. Microstructure and mechanical properties of plasma sprayed HA/YSZ/Ti-6Al-4V composite coatings. *Biomaterials* 2004;25:4009-4017.

[860] Gu YW, Khor KA, Pan D, Cheang P. Activity of plasma sprayed yttria stabilized zirconia reinforced hydroxyapatite/Ti-6Al-4V composite coatings in simulated body fluid. *Biomaterials* 2004;25:3177-3185.

[861] Best SM, Porter AE, Thian ES, Huang J. Bioceramics: past, present and for the future. *J. Eur. Ceram. Soc.* 2008;28:1319-1327.

[862] de Aza PN, Guitián F, de Aza S. Bioeutectic: a new ceramic material for human bone replacement. *Biomaterials* 1997;18:1285-1291.

[863] Huang X, Jiang D, Tan S. Apatite formation on the surface of wollastonite/tricalcium phosphate composite immersed in simulated body fluid. *J. Biomed. Mater. Res. B Appl. Biomater.* 2004;69B:70-72.

[864] Zhang F, Chang J, Lin K, Lu J. Preparation, mechanical properties and *in vitro* degradability of wollastonite/tricalcium phosphate macroporous

scaffolds from nanocomposite powders. *J. Mater. Sci. Mater. Med.* 2008;19:167-173.

[865] Juhasz JA, Best SM, Kawashita M, Miyata N, Kokubo T, Nakamura T, Bonfield W. Bonding strength of the apatite layer formed on glass-ceramic apatite-wollastonite-polyethylene composites. *J. Biomed. Mater. Res. A* 2003;67A:952-959.

[866] Juhasz JA, Best SM, Bonfield W, Kawashita M, Miyata N, Kokubo T, Nakamura T. Apatite-forming ability of glass-ceramic apatite-wollastonite – polyethylene composites: effect of filler content. *J. Mater. Sci. Mater. Med.* 2003;14:489-495.

[867] Juhasz JA, Best SM, Brooks R, Kawashita M, Miyata N, Kokubo T, Nakamura T, Bonfield W. Mechanical properties of glass-ceramic A-W-polyethylene composites: effect of filler content and particle size. *Biomaterials* 2004;25:949-955.

[868] Rea SM, Brooks RA, Best SM, Kokubo T, Bonfield W. Proliferation and differentiation of osteoblast-like cells on apatite-wollastonite/polyethylene composites. *Biomaterials* 2004;25:4503-4512.

[869] Greish YE, Brown PW. Characterization of wollastonite-reinforced HAp-Ca polycarboxylate composites. *J. Biomed. Mater. Res.* 2001;55:618-628. Erratum in: *J. Biomed. Mater. Res.* 2001;56:459.

[870] Greish YE, Brown PW. Characterization of bioactive glass-reinforced HAP-polymer composites. *J. Biomed. Mater. Res.* 2000;52:687-694.

[871] Kangasniemi I, de Groot K, Wolke J, Andersson O, Luklinska Z, Becht JGM, Lakkisto M, Yli-Urpo A. The stability of hydroxyapatite in an optimized bioactive glass matrix at sintering temperatures. *J. Mater. Sci. Mater. Med.* 1991;2:133-137.

[872] Kangasniemi IMO, de Groot K, Becht JGM, Yli-Urpo A. Preparation of dense hydroxylapatite or rhenanite containing bioactive glass composites. *J. Biomed. Mater. Res.* 1992;26:663-674.

[873] Kangasniemi IMO, Vedel E, de Blick-Hogerworst J, Yli-Urpo A, de Groot K. Dissolution and scanning electron microscopic studies of Ca,P particle-containing bioactive glasses. *J. Biomed. Mater. Res.* 1993;27:1225-1233.

[874] Maruno S, Ban S, Wang YF, Iwata H, Itoh H. Properties of functionally gradient composite consisting of hydroxyapatite containing glass coated titanium and characters for bioactive implant. *J. Ceram. Soc. Japan* 1992;100:362-367.

[875] White AA, Best SM, Kinloch IA. Hydroxyapatite-carbon nanotube composites for biomedical applications: a review. *Int. J. Appl. Ceram. Technol.* 2007;4:1-13.

[876] Chlopek J, Czajkowska B, Szaraniec B, Frackowiak E, Szostak K, Beguin F. In vitro studies of carbon nanotubes biocompatibility. *Carbon* 2006;44:1106-1111.

[877] Price RL, Waid MC, Haberstroh KM, Webster TJ. Selective bone cell adhesion on formulations containing carbon nanofibers. *Biomaterials* 2003;24:1877-1887.

[878] Zanello LP, Zhao B, Hu H, Haddon RC. Bone cell proliferation on carbon nanotubes. *Nano Lett.* 2006;6:562-567.

[879] Saito N, Usui Y, Aoki K, Narita N, Shimizu M, Ogiwara N, Nakamura K, Ishigaki N, Kato H, Taruta S. Carbon nanotubes for biomaterials in contact with bone. *Current Medicinal Chemistry* 2008;15:523-527.

[880] Kobayashi S, Kawai W. Development of carbon nanofiber reinforced hydroxyapatite with enhanced mechanical properties. *Composites A* 2007;38:114-123.

[881] Balani K, Anderson R, Laha T, Andara M, Tercero J, Crumpler E, Agarwal A. Plasma-sprayed carbon nanotube reinforced hydroxyapatite coatings and their interaction with human osteoblasts *in vitro*. *Biomaterials* 2007;28:618-624.

[882] Chen Y, Gan CH, Zhang TH, Yu G, Bai P, Kaplan A. Laser-surface-alloyed carbon nanotubes reinforced hydroxyapatite composite coatings. *Appl. Phys. Lett.* 2005;86:251905 (3 pages).

[883] Chen Y, Zhang TH, Gan CH, Yu G. Wear studies of hydroxyapatite composite coating reinforced by carbon nanotubes. *Carbon* 2007;45:998-1004.

[884] Chen Y, Zhang YQ, Zhang TH, Gan CH, Zheng CY, Yu G. Carbon nanotube reinforced hydroxyapatite composite coatings produced through laser surface alloying. *Carbon* 2006;44:37-45.

[885] Ding Y, Liu J, Jin X, Lu H, Shen G, Yu R. Poly-L-lysine/hydroxyapatite/carbon nanotube hybrid nanocomposite applied for piezoelectric immunoassay of carbohydrate antigen 19-9. *Analyst* 2008;133:184-190.

[886] Slosarcyk A, Klisch M, Blazewicz M, Piekarczyk J, Stobierski L, Rapacz-Kmita A. Hot pressed hydroxyapatite-carbon fibre composites. *J. Eur. Ceram. Soc.* 2000;20:1397-1402.

[887] Dorner-Reisel A, Berroth K, Neubauer R, Nestler K, Marx G, Scislo M, Müller E, Slosarcyk A. Unreinforced and carbon fibre reinforced

hydroxyapatite: resistance against microabrasion. *J. Eur. Ceram. Soc.* 2004;24:2131-2139.
[888] Fu T, Zhao JL, Wei JH, Han Y, Xu KW. Preparation of carbon fiber fabric reinforced hydroxyapatite/epoxy composite by RTM processing. *J. Mater. Sci.* 2004;39:1411-1413.
[889] Yoshimura M. Phase stability of zirconia. *Am. Ceram. Soc. Bull.* 1988;67:1950-1955.
[890] Thompson I, Rawlings RD. Mechanical behaviour of zirconia and zirconia-toughened alumina in a simulated body environment. *Biomaterials* 1990;11:505-508.
[891] Monma H. Tricalcium phosphate ceramics complexed with hydroxyapatite. *J. Ceram. Soc. Jpn.* 1987;96:60-64.
[892] Suchanek W, Yashima M, Kakihana M, Yoshimura M. Processing and mechanical properties of hydroxyapatite reinforced with hydroxyapatite whiskers. *Biomaterials* 1996;17:1715-1723.
[893] Suchanek W, Yashima M, Kakihana M, Yoshimura M. Hydroxyapatite/hydroxyapatite-whisker composites without sintering additives: mechanical properties and microstructural evolution. *J. Am. Ceram. Soc.* 1997;80:2805-2813.
[894] Kaito T, Mukai Y, Nishikawa M, Ando W, Yoshikawa H, Myoui A. Dual hydroxyapatite composite with porous and solid parts: Experimental study using canine lumbar interbody fusion model. *J. Biomed. Mater. Res. B Appl. Biomater.* 2006;78B:378-384.
[895] Ramay HR, Zhang M. Biphasic calcium phosphate nanocomposite porous scaffolds for load-bearing bone tissue engineering. *Biomaterials* 2004;21:5171-5180.
[896] Watari F, Yokoyama A, Saso F, Uo M, Kawasaki T. Fabrication and properties of functionally graded dental implant. *Composites Part B* 1997;28B:5-11.
[897] Watari F, Yokoyama A, Omori M, Hirai T. Kondo H, Uo M, Kawasaki T. Biocompatibility of materials and development to functionally graded implant for bio-medical application. *Compos. Sci. Technol.* 2004;64:893-908.
[898] Chenglim C, Jingchuan Z, Zhongda Y, Shidong W. Hydroxyapatite-Ti functionally graded biomaterial fabricated by powder metallurgy. *Mater. Sci. Eng. A 1999*;271:95-100.
[899] Inagaki M, Yokogawa Y, Kameyama T. Effects of plasma gas composition on bond strength of hydroxyapatite/titanium composite

coatings prepared by rf-plasma spraying. *J. Eur. Ceram. Soc.* 2006;26:495-499.

[900] Ban S, Hasegawa J, Maruno S. Fabrication and properties of functionally gradient bioactive composites comprising hydroxyapatite containing glass coated titanium. *Mater. Sci. Forum* 1999;308-311:350-355.

[901] Stojanovic D, Jokic B, Veljovic Dj, Petrovic R, Uskokovic PS, Janackovic Dj. Bioactive glass-apatite composite coating for titanium implant synthesized by electrophoretic deposition. *J. Eur. Ceram. Soc.* 2007;27:1595-1599.

[902] Nonami T, Kamiya A, Naganuma K, Kameyana T. Implantation of hydroxyapatite granules into superplastic titanium alloy for biomaterials. *Mater. Sci. Eng. C* 1998;6:281-284.

[903] Wong LH, Tio B, Miao X. Functionally graded tricalcium phosphate/fluoroapatite composites. *Mater. Sci. Eng. C* 2002;20:111-115.

[904] Peltola T, Patsi M, Rahiala H, Kangasniemi I, Yli-Urpo A. Calcium phosphate induction by sol-gel-derived titania coatings on titanium substrates *in vitro*. *J. Biomed. Mater. Res.* 1998;41:504-510.

[905] Heilmann F, Standard OC, Müller FA, Hoffman M. Development of graded hydroxyapatite/$CaCO_3$ composite structures for bone ingrowth. *J. Mater. Sci. Mater. Med.* 2007;18:1817-1824.

[906] Cavalcanti A, Shirinzadeh B, Zhang M, Kretly LC. Nanorobot hardware architecture for medical defense. *Sensors* 2008;8:2932-2958.

[907] Wypych G. *Handbook of fillers*. 2nd Ed. ChemTec Publishing, New York, 1999.

[908] Rhee SH, Lee JD, Tanaka J. Nucleation of hydroxyapatite crystal through chemical interaction with collagen. *J. Am. Ceram. Soc.* 2000;83:2890-2892.

[909] Lin X, Li X, Fan H, Wen X, Lu J, Zhang X. *In situ* synthesis of bone-like apatite/collagen nano-composite at low temperature. *Mater. Lett.* 2004;58:3569-3572.

[910] Zhang W, Liao SS, Cui FZ. Hierarchical self-assembly of nanofibrils in mineralized collagen. *Chem. Mater.* 2003;15:3221-3226.

[911] Liu Q, de Wijn JR, van Blitterswijk CA. Covalent bonding of PMMA, PBMA and poly(HEMA) to hydroxyapatite particles. *J. Biomed. Mater. Res.* 1998;40:257-263.

[912] Li J, Chen YP, Yin Y, Yao F, Yao K. Modulation of nano-hydroxyapatite size via formation on chitosan-gelatin network film *in situ*. *Biomaterials* 2007;28:781-790.
[913] Boanini E, Gazzano M, Rubini K, Bigi A. Composite nanocrystals provide new insight on alendronate interaction with hydroxyapatite structure. *Adv. Mater.* 2007;19:2499-2502.
[914] Tjandra W, Yao J, Ravi P, Tam KC, Alamsjah A. Nanotemplating of calcium phosphate using a double-hydrophilic block copolymer. *Chem. Mater.* 2005;17:4865-4872.
[915] Misra DN. Adsorption of zirconyl salts and their acids on hydroxyapatite: use of salts as coupling agents to dental polymer composites. *J. Dent. Res.* 1985;12:1405-1408.
[916] Liu Q, de Wijn JR, van Blitterswijk CA. A study on the grafting reaction of isocyanates with hydroxyapatite particles. *J. Biomed. Mater. Res.* 1998;40:358-364.
[917] Grossman RF. in: *Plastics additives and modifiers handbook*. 2nd ed. Edenbaum J. (Ed.), Chapman & Hall, New York, 1996.
[918] Chang MC, Ikoma T, Kikuchi M, Tanaka J. Preparation of a porous hydroxyapatite/collagen nanocomposite using glutataldehyde as a crosslinkage agent. *J. Mater. Sci. Lett.* 2001;20:1199-1201.
[919] Sousa RA, Reis RL, Cunha AM, Bevis MJ. Coupling of HDPE/hydroxyapatite composites by silane-based methodologies. *J. Mater. Sci. Mater. Med.* 2003;14:475-487.
[920] Wang M, Deb S, Bonfield W. Chemically coupled hydroxyapatite-polyethylene composites: processing and characterisation. *Mater. Lett.* 2000;44:119-124.
[921] Wang M, Bonfield W. Chemically coupled hydroxyapatite-polyethylene composites: structure and properties. *Biomaterials* 2001;22:1311-1320.
[922] Dupraz AMP, de Wijn JR, van der Meer SAT, Goedemoed JH. Biocompatibility screening of silane-treated hydroxyapatite powders, for use as filler in resorbable composites. *J. Mater. Sci. Mater. Med.* 1996;7:731-738.
[923] Dupraz AMP, de Wijn JR, van der Meer SAT, de Groot K. Characterization of silane-treated hydroxyapatite powders reinforced for use as filler in biodegradable composites. *J. Biomed. Mater. Res.* 1996;30:231-238.
[924] Liao JG, Wang XJ, Zuo Y, Zhang L, Wen JQ, Li Y. Surface modification of nano-hydroxyapatite with silane agent. *J. Inorg. Mater.* 2008;23:145-149.

[925] Sousa RA, Reis RL, Cunha AM, Bevis MJ. Structure development and interfacial interactions in HDPE/HA composites moulded with preferred orientation. *J. Appl. Polym. Sci.* 2002;86:2866-2872.
[926] Morita S, Furuya K, Ishihara K, Nakabayashi N. Performance of adhesive bone cement containing hydroxyapatite particles. *Biomaterials* 1998;19:1601-1606.
[927] Shinzato S, Nakamura T, Tamura J, Kokubo T, Kitamura Y. Bioactive bone cement: effects of phosphoric ester monomer on mechanical properties and osteoconductivity. *J. Biomed. Mater. Res.* 2001;56:571-577.
[928] Dorozhkin SV. Is there a chemical interaction between calcium phosphates and hydroxypropylmethylcellulose (HPMC) in organic/inorganic composites? *J. Biomed. Mater. Res.* 2001;54:247-255.
[929] Omori M, Okubo A, Otsubo M, Hashida T, Tohji K. Consolidation of multi-walled carbon nanotube and hydroxyapatite coating by the spark plasma system (SPS). *Key Engin. Mater.* 2004;254-256:395-398.
[930] Zhao B, Hu H, Mandal SK, Haddon RC. A bone mimic based on the self-assembly of hydroxyapatite on chemically functionalized single-walled carbon nanotubes. *Chem. Mater.* 2005;17:3235-3241.
[931] Kasuga T, Yoshida M, Ikushima AJ, Tuchiya M, Kusakari H. Bioactivity of zirconia-toughened glass-ceramics. *J. Am. Ceram. Soc.* 1992;75:1884-1888.
[932] Dorozhkin SV. Inorganic chemistry of the dissolution phenomenon: the dissolution mechanism of calcium apatites at the atomic (ionic) level. *Comments Inorg. Chem.* 1999;20:285-299.
[933] Dorozhkin SV. A review on the dissolution models of calcium apatites. *Prog. Crystal Growth Charact.* 2002;44:45-61.
[934] Furukawa T, Matsusue Y, Yasunaga T, Shikinami Y, Okuno M, Nakamura T. Biodegradation behavior of ultra-high strength hydroxyapatite/poly(L-lactide) composite rods for internal fixation of bone fractures. *Biomaterials* 2000;21:889-898.
[935] Furukawa T, Matsusue Y, Yasunaga T, Nakagawa Y, Okada Y, Shikinami Y, Okuno M, Nakamura T. Histomorphometric study on high-strength hydroxyapatite/poly(L-lactide) composite rods for internal fixation of bone fractures. *J. Biomed. Mater. Res.* 2000;50:410-419.
[936] Yasunaga T, Matsusue Y, Furukawa T, Shikinami Y, Okuno M, Nakamura T. Bonding behaviour of ultrahigh strength unsintered hydroxyapatite particles/poly(L-lactide) composites to surface of tibial cortex in rabbits. *J. Biomed. Mater. Res.* 1999;47:412-419.

[937] Marques AP, Reis RL, Hunt JA. *In vitro* evaluation of the biocompatibility of novel starch based polymeric and composite material. *Biomaterials* 2002;6:1471-1478.

[938] Mendes SC, Bovell YP, Reis RL, Cunha AM, de Bruijn JD, van Blitterswijk CA. Biocompatibility testing of novel starch-based materials with potential application in orthopaedic surgery. *Biomaterials* 2001;22:2057-2064.

[939] Meyers MA, Lin AYM, Seki Y, Chen PY, Kad BK, Bodde S. Structural biological composites: an overview. *JOM* 2006;58:36-43.

[940] Dorozhkin, S.V. Nano-sized and nanocrystalline calcium orhophosphates in biomedical engineering. *J. Biomimetics, Biomaterials and Tissue Engineering* 2009, 3, 59-92.

Chapter 10

NOMENCLATURE

EVOH	ethylene-vinyl alcohol copolymer
IBS	injectable bone substitute
HDPE	high-density polyethylene
HIPS	high impact polystyrene
HPMC	hydroxypropylmethylcellulose
PAA	polyacrylic acid
PBT	polybutyleneterephthalate
PCL	poly(ε-caprolactone)
PDLLA	poly-DL-lactic acid
PE	polyethylene
PEEK	polyetheretherketone
PEG	polyethylene glycol
PGA	polyglycolic acid
PHB	polyhydroxybutyrate
PHBHV	poly(hydroxybutyrate-*co*-hydroxyvalerate)
PHEMA	polyhydroxyethyl methacrylate
PHV	polyhydroxyvalerate
PLA	polylactic acid
PLGA	poly(lactic-*co*-glycolic) acid
PLGC	co-polyester lactide-*co*-glycolide-*co*-ε-caprolactone
PLLA	poly(L-lactic acid)
PMMA	polymethylmethacrylate
PPF	poly(propylene-*co*-fumarate)
PS	polysulfone
PSZ	partially stabilized zirconia
PTFE	polytetrafluoroethylene
PVA	polyvinyl alcohol

PVAP	polyvinyl alcohol phosphate
SEVA	ethylene vinyl alcohol copolymer
UHMWPE	ultrahigh molecular weight polyethylene

INDEX

A

absorption, 25, 58, 110
absorption spectra, 58
accelerator, 85
acetic acid, 41
acid, 3, 15, 16, 23, 25, 28, 38, 41, 45, 46, 62, 82, 83, 84, 88, 92, 93, 96, 98, 99, 106, 112, 113, 116, 119, 120, 121, 122, 124, 130, 132, 153
acidic, 22, 25, 31, 42, 65, 132
acrylic acid, 62, 119, 122
acrylonitrile, 13
activation, 56
active site, 58
acute, 42, 127
adaptation, 78
additives, 18, 25, 34, 111, 148, 150
adhesion, 11, 14, 23, 37, 43, 44, 45, 61, 62, 63, 67, 80, 98, 100, 110, 113, 119, 128, 130, 134, 141, 147
adhesive properties, 128
adsorption, 55, 94
agent, 23, 24, 32, 46, 61, 62, 110, 115, 119, 150
agents, 42, 44, 47, 61, 62, 63, 83, 101, 139, 150
aggregates, 34, 63, 103
aging, 1
aid, 41
albumin, 33
alcohol, 67, 93, 95, 96, 117, 118, 128, 132, 153, 154
alendronate, 59, 150
alkali, 134
alkaline, 62, 65
allografts, 1
alloys, 16, 48
allylamine, 46
alternative, 63, 73
alternatives, 4
aluminates, 64
aluminium, 140
amino, 3, 14, 23, 41, 46, 59, 63, 93, 132
amino acid, 3, 14, 23, 41, 46, 93, 132
amino groups, 59, 63
amorphous, 15, 17, 28, 102, 108, 113, 135
amorphous carbon, 28, 113
Amsterdam, 81, 82
angiogenesis, 71
annealing, 91
anther, 121
antibiotic, 29, 44, 128, 129, 130
antibiotics, 145
antigen, 56, 147
Apatite, 24, 41, 49, 86, 125, 143, 145, 146
apatite layer, 146
apoptosis, 37, 94
application, viii, 1, 4, 5, 11, 14, 16, 30, 46, 47, 48, 53, 56, 60, 62, 69, 70, 73, 80, 88, 93, 103, 104, 143, 148, 152
appropriate technology, 44
aqueous solution, 12, 24, 26, 35, 41, 47, 63

156   Index

aqueous solutions, 12, 24, 26, 35, 41, 63
aqueous suspension, 41
arthroplasty, 82
aspect ratio, 20, 49, 122
assessment, 70, 88, 102, 106, 122
atmosphere, 29
atomic force, 101
atomic force microscopy, 101
atoms, 33, 37, 59, 60
atrophy, 65
attachment, 20, 29, 31, 39, 41, 58, 90, 91
autografts, 1, 42
autologous bone, 75

B

bacteria, 44, 79
bacterial, 1, 79
barium, 48
batteries, 33
battery, 33, 56
behavior, 4, 5, 8, 11, 21, 22, 25, 28, 31, 32, 44, 58, 61, 67, 70, 73, 90, 92, 96, 97, 98, 110, 113, 143, 151
bending, 17, 25, 27, 49, 51
benign, 4
bile, 132
bile duct, 132
binding, 56, 60
binding energies, 60
bioactive materials, 76
biocompatibility test, 70
biocompatible, viii, 1, 2, 4, 11, 13, 14, 16, 17, 22, 28, 31, 33, 38, 46, 47, 52, 54, 79, 96, 102, 142
biodegradability, 45, 65
biodegradable materials, 65, 82, 94
biodegradation, 27, 28, 31, 43, 67, 105
Bioglass, 17, 48, 87, 136
bioinert, 17, 21, 48, 90
biological macromolecules, 19, 44, 56
biological systems, 73
biomaterial, 4, 17, 29, 52, 77, 87, 100, 107, 114, 118, 124, 125, 127, 135, 144, 148

biomaterials, vii, viii, 2, 4, 5, 11, 17, 19, 20, 34, 44, 49, 55, 57, 59, 66, 69, 70, 71, 73, 74, 75, 76, 77, 78, 79, 80, 83, 84, 85, 92, 96, 100, 101, 102, 103, 107, 125, 127, 135, 147, 149
biomechanics, 77
biomedical applications, vii, 11, 13, 16, 17, 51, 69, 73, 79, 82, 90, 91, 112, 136, 138, 147
biomimetic, 17, 41, 43, 44, 51, 80, 92, 93, 95, 104, 109, 116, 117, 118, 121, 123, 125, 126, 127, 129, 130, 143
biomineralization, 58, 70, 76
biomolecules, 55, 70
biopolymer, 85
biosensors, 19, 56
bis-phenol, 35
blends, 14, 22, 23, 28, 100, 101
blocks, 24, 28, 29, 88, 107
blood, 1, 3, 73
blood group, 1
blood vessels, 3
body fluid, 25, 31, 43, 79, 93, 131, 144, 145
bonding, 2, 53, 57, 58, 59, 60, 62, 63, 65, 67, 83, 92, 118, 120, 140, 143, 144, 149
bonds, 17, 34, 58, 59, 61
bone cement, 14, 82, 109, 110, 111, 112, 126, 142, 144, 151
bone graft, vii, 1, 4, 28, 33, 42, 43, 44, 46, 51, 69, 70, 73, 75, 76, 77, 85, 107, 117, 124
bone grafts, 1, 42, 44, 46, 51, 69, 70, 73
bone growth, 1, 4, 70
bone marrow, 39, 42, 134, 136, 143
bone morphogenetic proteins, 84
bone powder, 84
bone remodeling, 44, 109
Boston, 81
bovine, 1, 2, 42, 45, 125, 127, 131, 142
breakdown, 13
building blocks, 93
by-products, 65

C

calcium carbonate, 99
calvaria, 39, 122
candidates, vii
caprolactone, 16, 28, 78, 102, 103, 105, 106, 107, 119, 120, 136, 153
carbide, 48, 137
carbohydrate, 147
carbon, viii, 23, 34, 47, 49, 50, 55, 64, 80, 92, 93, 94, 96, 109, 147, 148, 151
Carbon, 16, 49, 93, 95, 147
carbon dioxide, 47, 94
carbon nanotubes, viii, 23, 34, 49, 50, 64, 93, 147, 151
Carbon nanotubes, 49, 93, 147
carbonates, 3
carboxyl, 50, 59, 63
carboxyl groups, 63
carboxymethyl cellulose, 121
caries, 38
carrier, 44, 47, 84, 117, 124, 145
cartilage, 28, 83, 92, 125
casein, 46, 135
cast, 63
casting, 23, 26, 53, 137, 142
cavities, 62, 67
cell, 11, 13, 15, 20, 22, 23, 25, 28, 29, 31, 39, 43, 44, 45, 67, 73, 76, 81, 90, 98, 107, 113, 126, 127, 128, 130, 134, 138, 147
cell adhesion, 67, 128, 130, 147
cell culture, 25, 28, 29, 39, 43, 45, 107
cell growth, 126, 127
cellulose, 16, 79, 83
Cellulose, 14, 83
cement, 19, 32, 34, 35, 36, 63, 108, 109, 110, 111, 112, 113, 126, 131, 142
ceramic, 2, 3, 20, 22, 34, 36, 41, 45, 48, 61, 80, 87, 88, 89, 90, 92, 93, 134, 137, 138, 142, 143, 145, 146
channels, 13
chemical composition, 2, 4, 13, 17, 37, 42, 48
chemical interaction, 34, 59, 63, 149, 151

chemical properties, 7
chemical stability, 29, 67
chemicals, 25, 62
China, 120
chitin, vii, 16, 17, 45, 84, 88, 106, 116, 131
chitosan, 16, 31, 33, 36, 38, 39, 45, 46, 60, 63, 84, 85, 92, 96, 106, 108, 109, 112, 119, 120, 121, 122, 129, 130, 131, 132, 134, 150
Chitosan, 17, 31, 36, 85, 131
chloride, 133
chloroform, 28
ciprofloxacin, 128
classes, 2
classification, 8, 74
cleaning, 70
clinical trial, 127, 136
clinics, 74, 77
clusters, 47
coagulation, 46
coatings, 7, 11, 13, 24, 50, 53, 78, 79, 80, 92, 94, 98, 102, 125, 131, 136, 138, 139, 141, 142, 145, 147, 149
cocoon, 46
collagen, vii, viii, 3, 16, 19, 22, 29, 32, 34, 37, 38, 39, 40, 41, 42, 43, 44, 53, 58, 61, 67, 77, 92, 103, 107, 109, 115, 116, 117, 121, 122, 123, 124, 125, 126, 127, 128, 149, 150
colonization, 11, 46, 67
community, 34
compaction, 23, 140
compatibility, 9, 25
compensation, 98
complexity, 104
components, 3, 5, 7, 13, 20, 22, 24, 34, 40, 62, 64, 67, 70, 73
composition, 1, 2, 3, 4, 5, 13, 17, 37, 42, 48, 49, 52, 54, 73, 90, 91, 103, 120, 141, 148
compounds, 7, 12, 19, 24, 25, 44, 48, 59, 61, 99
compressive strength, 22, 23, 35, 50, 52, 111
computed tomography, 13
concentration, 24, 29, 46, 47, 54, 56

concrete, 35
conduction, 108
conductive, 21
conductive hearing loss, 21
configuration, 38
Congress, iv
connective tissue, 17
conservation, 1
constituent materials, 7
construction, 5, 34, 36
control, 34, 41, 44, 54, 61, 70, 80, 99, 100, 136
convergence, 53
cooling, 21, 57
copolymer, 14, 38, 60, 67, 88, 92, 96, 101, 106, 120, 128, 150, 153, 154
copolymerization, 13
copolymers, 14, 16, 23, 45, 62, 63, 84, 92
coral, 2
core-shell, 33, 38, 108
cornea, 24, 95
correlation, 60
corrosion, 4, 17
cortex, 151
coupling, 23, 24, 61, 62, 101, 110, 119, 150
covalent, 57, 58, 61, 64, 120
covalent bond, 57, 58, 61
crack, 8, 34
cracking, 18
craniofacial, 133
cranioplasty, 110
dental implants, 52, 80, 83
dentistry, 78
deposition, 22, 37, 41, 58, 106, 141, 149
derivatives, 16
detection, 55, 144
differential scanning, 60
differential scanning calorimetry, 60
differentiation, 37, 39, 67, 105, 126, 134, 143, 146
diffusion, 30, 64, 104
dimethacrylate, 111
dipole, 104
diseases, 42
dispersion, 8, 20, 22, 34, 38, 101

distribution, 8, 21, 52, 69, 103
dogs, 126
donor, 1
dopants, 24
doped, 142
drug delivery, 13, 32, 33, 78, 81, 106, 107, 117, 131
drug delivery systems, 106
drug release, 29, 102, 107, 129, 130
drugs, 15, 31, 36
drying, 23, 26, 38, 43, 45, 46, 93
ductility, 13, 22
dynamic mechanical analysis, 60

E

elaboration, 74
elasticity, 2, 3, 37, 128
elastin, 124, 127
electric current, 145
electric field, 103
electric power, 56
Electroanalysis, 119
electron, 61, 63, 86, 146
electron microscopy, 61, 63, 64
electronic structure, 37
electrospinning, 120, 143
elongation, 114
embryonic stem, 98
emulsions, 23
encapsulated, 2
encapsulation, 13, 24
energy, 8, 23, 33, 58, 62, 64, 94, 123
environment, 4, 13, 40, 86, 148
enzymatic, 14, 123
epitaxial growth, 21
epithelium, 128
epoxy, 148
equilibrium, 29
erosion, 8, 14
ester, 63, 78, 151
esters, 14, 84
ethylene, 13, 38, 67, 106, 107, 111, 119, 121, 128, 132, 153, 154
ethylene glycol, 106, 107, 111, 121, 132

ethylene oxide, 13
ethylene vinyl alcohol, 67, 154
evaporation, 28
evolution, 73, 95, 112, 148
exoskeleton, 17
external fixation, 42
extraction, 2
extrusion, 22, 89, 142

F

FA, 12, 24, 26, 29, 30, 37, 53, 55, 60, 149
fabric, 148
fabricate, 33, 36, 44, 52, 70, 93
fabrication, vii, 5, 7, 9, 23, 24, 47, 67, 69, 70, 87, 132, 134, 135, 140
factorial, 65
failure, 1, 21, 89
family, 45
fatigue, 11, 34, 70, 89, 97, 110
FDA, 14, 15, 20
FDA approval, 14, 20
femoral bone, 124
femur, 4, 36, 43, 106
fiber, 40, 57, 78, 89, 108, 126, 131, 148
fibers, 3, 5, 8, 16, 20, 28, 33, 37, 40, 41, 43, 48, 50, 57, 59, 62, 88, 99, 120, 121, 123
fibrillar, 41
fibrils, 39, 40, 41, 126
fibrin, 16, 46, 75, 133, 134
Fibrin, 46, 133
fibroblast, 33, 43, 46, 90, 107
fibroblasts, 98
fibronectin, 46
fibrous tissue, 2, 27, 67
filler particles, 23, 57, 62, 92
fillers, vii, 5, 11, 22, 28, 35, 43, 48, 61, 62, 63, 100, 114, 149
film, 150
films, 15, 29, 58, 131
fixation, 14, 35
flame, 23, 94
flexibility, 49
flow, 48, 86
fluid, 34, 53, 79, 86

Fluorapatite, 12, 104
fluoride, 30, 104, 139
fluorine, 138
foams, 25, 44, 78, 93, 94, 123
Fourier, 58
fractal growth, 30, 103
fracture, 1, 2, 4, 5, 17, 18, 21, 23, 41, 42, 49, 51, 62, 77, 78, 94, 122, 127, 142
fractures, 3, 14, 15, 42, 127, 151
fragility, 53
friction, 18, 50
fuel, 141
fumarate, 82, 99, 104, 108, 111, 117, 121, 153
fumaric, 82
functionalization, 50, 64
fusion, 51, 122, 124, 148

G

gas, 24, 148
gel, 23, 30, 43, 48, 98, 111, 118, 123, 124, 142
gelatin, vii, 16, 30, 31, 32, 34, 35, 38, 44, 48, 60, 93, 95, 103, 104, 105, 106, 107, 109, 111, 117, 129, 130, 150
gels, 13, 46, 47, 96
gene, 97, 122, 132
gene expression, 122
gene therapy, 97, 132
general knowledge, 20
generation, 74, 78, 129
gentamicin, 128, 129
geochemical, 81
geochemistry, 81
Germany, 125
glass, 2, 16, 17, 36, 48, 49, 53, 87, 108, 112, 136, 142, 143, 144, 146, 149, 151
glass transition, 16
glass transition temperature, 16
glasses, 2, 17, 19, 48, 49, 66, 67, 86, 146
glutamic acid, 46, 132
glutaraldehyde, 32, 43, 44, 62, 105, 106, 115, 117, 122
glycol, 153

gold, 23, 55, 93, 145
gold nanoparticles, 23, 55, 93
government, iv
grafting, 2, 29, 63, 76, 88, 150
grafting reaction, 150
grafts, 2, 40, 70, 77
grain, 37, 42, 63
grain boundaries, 37, 63
granules, 8, 27, 35, 40, 53, 63, 87, 133, 136, 149
groups, 14, 24, 25, 31, 59, 61, 62, 63
growth, 1, 4, 8, 15, 21, 24, 30, 32, 36, 38, 40, 41, 42, 43, 70, 71, 80, 97, 103, 114, 123, 125, 126, 127, 128, 129, 131, 142
growth factor, 15, 36, 38, 43
growth factors, 15, 36, 38, 43
growth mechanism, 129

H

hafnium, 86
handling, 27
hard tissues, 40, 46, 88
hardness, 3, 37, 50, 142
harvest, 42
harvesting, 1, 51, 75
healing, 2, 4, 9, 31, 65, 73, 78, 85, 127
health, 50
health problems, 50
hearing, 21
heat, 14, 17, 35, 49, 95, 140
helix, 3
herbal, 105
heterogeneous, 7
high pressure, 23
high temperature, 51
high-frequency, 140
hip, 14, 82
hip replacement, 14, 82
histological, 51, 101, 130
histology, 78
HK, 96
homogeneity, 8, 26
hospital, vii
host, 4, 24, 42

host tissue, 4, 24
hot pressing, 26, 51, 137
human, 1, 2, 4, 9, 17, 22, 25, 28, 29, 44, 46, 49, 50, 80, 84, 90, 91, 98, 102, 106, 126, 127, 129, 134, 136, 141, 145, 147
human mesenchymal stem cells, 134
humans, 73
hybrid, vii, viii, 5, 9, 19, 20, 31, 33, 38, 50, 55, 57, 59, 69, 70, 71, 74, 77, 80, 94, 98, 102, 105, 108, 119, 132, 141, 145, 147
hybridization, 122
hybrids, 95, 119
hydration, 25, 34
hydro, 32, 135, 150
hydrocarbon, 59
hydrogels, 45, 83, 121, 132
hydrogen, 57, 58, 59, 60, 118
hydrogen bonds, 58, 59
hydrolysis, 14, 26, 63
hydrophilic, 32, 135, 150
Hydrophilic, 107
hydrophilicity, 25, 29
hydrophobic, 22, 58
hydrophobic interactions, 58
hydrophobicity, 13, 67
hydrothermal, 141
Hydrothermal, 138
hydroxyacids, 83, 84
hydroxyl, 59, 78, 94
hydroxylapatite, 79, 83, 98, 100, 101, 127, 139, 140, 146

I

images, 63
imaging, 13
immersion, 31, 34, 110
immobilization, 56, 118
immunogenicity, 1
immunological, 1
implants, 5, 11, 14, 16, 21, 24, 25, 28, 34, 52, 53, 70, 76, 79, 80, 83, 84, 86, 95, 96, 97, 98, 99, 115, 124, 133, 140, 141
impregnation, 104
impurities, 16

in situ, 14, 21, 22, 28, 33, 35, 45, 63, 92, 93, 97, 101, 102, 122, 130, 131, 135, 141, 144, 150
in situ hybridization, 122
in vitro, vii, 25, 28, 31, 38, 43, 45, 49, 67, 76, 90, 94, 95, 99, 101, 102, 105, 107, 109, 110, 115, 116, 118, 119, 125, 129, 130, 134, 141, 142, 143, 145, 147, 149
in vivo, vii, 2, 5, 9, 13, 14, 24, 28, 31, 34, 36, 40, 41, 49, 50, 51, 67, 70, 71, 80, 84, 85, 87, 95, 101, 103, 105, 110, 112, 113, 115, 116, 117, 118, 122, 124, 126, 128, 133, 140
inclusion, 25, 45, 52
incompatibility, 1
incubation, 25
independence, 68
induction, 104, 125, 140, 149
industrial, 69, 81
industrial production, 69
infection, 44
infections, 1, 145
inflammatory, 13, 14, 25
inflammatory response, 13
infrared, 58, 62, 123
inhibition, 32
inhomogeneity, 9
injection, 20, 47, 91, 97, 100
injury, iv
inorganic, 2, 3, 5, 9, 19, 24, 26, 33, 37, 39, 40, 48, 64, 70, 80, 81, 85, 93, 98, 129, 131, 132, 151
inorganic filler, 98
inorganic fillers, 98
insight, 150
insulation, 13
integration, 24, 134
integrity, 47, 48
interaction, 5, 40, 55, 57, 58, 59, 61, 63, 64, 70, 103, 105, 110, 118, 129, 132, 134, 135, 147, 150
interactions, 22, 26, 29, 34, 40, 45, 58, 62, 63, 91, 101, 107, 118, 119, 123, 151
interface, 4, 17, 48, 53, 55, 58, 60, 62, 63, 78, 79, 80, 101, 124, 128, 135, 144

interfacial adhesion, 62, 119
interfacial bonding, 23, 42, 50, 58, 62, 70, 83
internal fixation, 122, 151
intrinsic, 40, 103
ionic, 12, 38, 41, 58, 61, 151
ions, 7, 17, 22, 31, 41, 45, 59, 70
IR spectroscopy, 84
irradiation, 13, 103, 139
irritation, 4
isostatic pressing, 51, 137
isotropic, 9

J

Japan, 105, 137, 139, 146
jaw, 43
Jordan, 85
Jun, 140
Jung, 126, 128

K

kinetics, 4, 25, 97

L

labor, 26
lactic acid, 25, 28, 83, 88, 92, 93, 98, 99, 106, 116, 153
laminin, 44, 128, 129
Laminin, 128
Langmuir, 93
laser, 50, 147
leaching, 24, 29, 48
leukemia, 37, 113
ligament, 78
limitation, 48
limitations, 1, 5, 11, 69
linear, 14, 21
linkage, 117, 120, 125, 129
lipids, 3, 13, 123
liquid phase, 31
lithography, 46

liver, 132
L-lactide, 26, 84, 91, 94, 95, 98, 99, 100, 103, 104, 105, 107, 114
London, 81
long period, 1
low temperatures, 112
lumbar, 67, 122, 148
lumbar spine, 67
lysine, 50, 55, 147

M

macromolecules, 19, 39, 44, 56, 59
macrophages, 98
macropores, 36
magnesium, 4, 17, 86, 87
magnesium alloys, 17
magnetic, iv, 13, 28
magnetic resonance, 13
magnetic resonance imaging, 13
maintenance, 78
mammals, 73
management, 42
mandibular, 31
mannitol, 36
manufacturing, 42, 55
mapping, 94
market, 2
marrow, 42, 105, 115
Marx, 147
matrix protein, 1
maxilla, 128
maxillary, 80
maxillary sinus, 80
measurement, 59, 63
mechanical behavior, 11, 21, 22, 25, 73, 90, 97, 98
mechanical testing, 31
media, 65, 83, 131, 133
mediation, 102
medicine, 14, 25, 69, 71, 76, 81
medulla, 52
melt, 17, 22, 26, 29, 52
melting, 16
membranes, 15, 92, 107, 125

mesenchymal stem cell, 105, 116, 130
mesenchymal stem cells, 105, 116, 134
metabolic, 14
metabolic pathways, 14
metabolism, 44
metallurgy, 51, 52, 144, 148
metals, viii, 8, 16, 19, 48, 86
methanol, 121
methyl methacrylate, 102
mice, 106
micelles, 33, 108
microemulsion, 26
micrometer, 70
microparticles, 27, 36, 47, 136
microscopy, 61, 63, 86, 101
microspheres, 26, 43, 93, 114, 127, 129
microstructure, 24, 115, 119, 130, 134, 139, 141, 144
microstructures, 44, 142
microwave, 139, 141
migration, 15, 23, 43, 84, 116
mimicking, 43, 79
mimicry, 79
mineralization, 23, 33, 38, 45, 94, 96, 108, 123, 126, 132
mineralized, 22, 40, 43, 77, 79, 91, 97, 109, 116, 123, 128, 149
mineralogy, 81
minerals, 37, 58
mixing, 26, 34, 41, 48, 58, 63
model system, 104
modeling, 80, 98, 108, 110
models, 61, 151
modulus, 15, 16, 20, 21, 22, 23, 25, 27, 28, 29, 31, 45, 49, 50, 51, 94
molar ratio, 53
molar ratios, 53
mold, 9
molecular mass, 91
molecular mechanisms, 71
molecular weight, 13, 21, 35, 47, 67, 82, 91, 103, 154
molecular weight distribution, 103
molecules, 13, 25, 39, 61, 63, 118
monomer, 28, 35, 67, 151

monomers, 13, 14, 15
morbidity, 2, 75
morphogenesis, 103, 104
morphological, 95, 132, 140
morphology, 67, 70, 78, 88, 103, 106, 110, 117
mouse, 46, 133
mouse model, 133
muscles, 134
musculoskeletal, 81

N

nanobiotechnology, 75
nanocapsules, 108
nanocomposites, 19, 37, 38, 50, 77, 79, 91, 96, 102, 104, 108, 114, 115, 117, 119, 120, 121, 122, 125, 126, 129, 130, 131, 132, 134, 139
Nanocomposites, 114, 132
nanocrystalline, 38, 42, 74, 152
nanocrystals, 3, 38, 39, 40, 59, 63, 96, 102, 114, 120, 121, 123, 132, 150
nano-crystals, 114
nanofibers, 120, 125, 147
nanoindentation, 50, 94
nanomaterials, 75, 94
nanometer, 37, 70
nanoparticles, 23, 38, 55, 93, 113, 115, 131, 132, 134, 140
nanorods, 38, 141
nanostructures, 33
nanotube, 50, 55, 80, 92, 109, 147, 151
nanotubes, viii, 23, 34, 49, 50, 64, 93, 147, 151
natural, 1, 2, 3, 16, 33, 37, 40, 41, 42, 43, 44, 46, 52, 70, 76, 104
needles, 47
network, 17, 32, 35, 40, 41, 82, 92, 99, 120, 129, 130, 134, 150
neutralization, 41
neutrophils, 116
New York, iii, iv, 76, 81, 149, 150
next generation, 5
nickel, 86
niobium, 86
nitrogen, 29
non-biological, 40
non-magnetic, 13
nontoxic, 13
norfloxacin, 128
nucleation, 24, 25, 40, 50, 59, 63
nylon, 96

O

observations, 61
occlusion, 85
oil, 33, 43, 45
optical, 37
optical properties, 37
oral, 81
organ, 85, 115
organic, 5, 37, 40, 46, 50, 63, 70, 77, 79, 80, 131, 132, 135, 151
organic solvent, 50
organic solvents, 50
orientation, 3, 8, 20, 21, 39, 60, 91, 100, 151
orthopaedic, 75, 79, 82, 85, 86, 98, 101, 139, 152
orthopedic surgeon, 1
Orthophosphate, 1, iii, 19, 32, 33, 37, 57, 65
orthophosphates, vii, viii, 2, 3, 4, 5, 7, 11, 12, 15, 17, 20, 21, 22, 25, 26, 27, 28, 31, 32, 34, 35, 37, 39, 40, 41, 43, 44, 45, 46, 48, 49, 51, 55, 57, 59, 61, 62, 64, 65, 66, 73, 76, 81, 108
osteoblasts, 3, 21, 23, 29, 31, 32, 43, 67, 89, 90, 91, 106, 110, 113, 125, 127, 141, 142, 147
osteocalcin, 3, 34, 80
osteoclastic, 80
osteoclasts, 3, 134
osteocytes, 3
osteoinductive, 1, 4, 31, 39, 40, 42, 69, 133, 145
osteomyelitis, 106
osteonectin, 3
osteopontin, 3
otitis media, 133

oxidation, 64
oxide, 13, 14, 18, 87, 140

P

pacemaker, 13
pain, 2
parameter, 38
particles, 5, 8, 20, 21, 23, 24, 25, 26, 28, 29, 31, 34, 41, 42, 45, 48, 49, 51, 57, 61, 62, 64, 80, 84, 87, 91, 92, 99, 114, 120, 121, 143, 144, 149, 150, 151
pathways, 14, 70
PEEK, 24, 97, 153
peptide, 127
peptides, 124
periodic, 104
periodontal, 119
permit, 4
pH, 12, 15, 25, 31, 33, 38, 41, 47, 62, 63, 98, 99, 129
pH values, 129
phase diagram, 136
phosphates, 81, 83, 107, 108, 130, 131, 151
physical interaction, 26, 57
physical properties, 9, 16, 112
physicochemical, 11, 55, 76, 94, 98, 134
physicochemical properties, 98, 134
physiological, 13, 46, 110, 126
piezoelectric, 42, 55, 144, 147
piezoelectric properties, 42
plasma, 50, 53, 80, 138, 139, 141, 142, 145, 148, 151
plastic, 48, 88, 133
plasticity, 9, 47
platelet, 137
platelets, 48
play, vii, 17
polar groups, 59
poly(3-hydroxybutyrate), 28
poly(lactic-*co*-glycolic acid), 98
poly(L-lactide), 78, 98, 114, 116, 151
polyacrylonitrile fiber, 121
polyamide, 38, 39, 59, 60, 91, 117, 118
polycrystalline, 18

polydimethylsiloxane, 35, 111
polydispersity, 13
polyester, 13, 31, 112, 153
polyesters, 13, 99
polyetheretherketone, 87, 97, 153
polyethylene, 14, 82, 88, 89, 90, 91, 146, 153, 154
polyglycolic acid, 15, 84, 153
polyhydroxybutyrate, 101, 104, 153
polymer blends, 92
polymer chains, 45
polymer composites, 33, 78, 89, 90, 92, 93, 94, 98, 108, 111, 112, 135, 146, 150
polymer matrix, 5, 24, 25
polymer nanocomposites, 16, 85
polymer properties, 13
polymeric composites, 90, 91, 108
polymeric materials, 13
polymeric matrices, 62
polymerization, 9, 14, 26, 28, 46, 63, 88, 92, 97, 102, 110, 114
polymerization process, 92
polymers, viii, 3, 5, 8, 9, 13, 14, 15, 16, 19, 20, 22, 23, 26, 28, 29, 35, 38, 47, 48, 61, 62, 67, 75, 82, 84, 85, 86, 97, 100, 103, 120
polymethylmethacrylate, 110, 111, 153
polyphosphazene, 35, 36, 120
polypropylene, 14, 89
polysaccharide, 17, 130
polysaccharides, 3, 13, 16
polystyrene, 153
polytetrafluoroethylene, 153
polyurethane, 32, 36
polyurethane foam, 32, 36
polyurethanes, 82
polyvinyl alcohol, 119, 153, 154
polyvinylalcohol, 95
poor, 2, 11, 14, 21, 22, 23, 42, 44, 45, 48, 73
population, 1
pore, 2, 32, 37, 42, 46, 67, 70, 86, 87
porosity, 4, 11, 17, 21, 23, 31, 32, 35, 36, 38, 42, 43, 46, 51, 52, 53, 67, 86, 106
porous metals, 16, 86
postoperative, 2

powder, 20, 22, 28, 34, 47, 48, 49, 51, 52, 109, 119, 126, 140, 143, 144, 148
powders, 45, 50, 62, 87, 93, 98, 137, 138, 144, 146, 150
power, 56, 124
precipitation, 22, 24, 27, 33, 38, 41, 42, 45, 51, 63, 92, 130, 131, 137, 140
preclinical, 82
pressure, 23, 80, 91
proactive, 84
probe, 61, 86
production, 17, 20, 41, 69, 98
proliferation, 11, 15, 20, 29, 37, 43, 67, 100, 130, 147
propagation, 8, 34
property, iv, 8, 50, 52, 53, 61, 83, 108
propylene, 82, 99, 104, 108, 111, 117, 153
prostheses, 18, 49, 82, 90
prosthesis, 1, 21, 139
protein, 39, 44, 46, 84, 94, 96, 106, 107, 117, 122, 124, 133
proteins, 1, 3, 13, 15, 16, 22, 44, 45, 55, 58, 84, 113
Proteins, 131
proteoglycans, 123
pseudo, 17
PT, 83, 89
PTFE, 13, 153
pulse, 145

R

radioactive waste, 145
radius, 38
Raman, 79, 123
Raman spectroscopy, 123
range, 3, 12, 13, 17, 21, 28, 35
rapid prototyping, 125
rat, 31, 106, 113, 128
rats, 31, 36, 39, 43, 88, 122
raw material, 17
raw materials, 17
reactivity, 9, 57
reagent, 62
reagents, 61

reconstruction, 13, 21, 75, 82, 99, 101, 113, 133
reconstructive surgery, 127
recovery, 77
red shift, 59
redistribution, 8
redox, 118
redox-active, 118
refractive index, 9
refractory, 86
regenerated cellulose, 83
regeneration, vii, 5, 14, 15, 22, 25, 28, 39, 47, 71, 77, 78, 85, 91, 93, 96, 97, 98, 99, 102, 104, 105, 116, 122, 125, 129, 135
regenerative medicine, 83
regular, 14, 58
reinforcement, 8, 20, 22, 24, 28, 31, 34, 48, 57, 62, 63, 69, 89, 100, 114
relationships, 83
relaxation, 53
reliability, 2, 48
remodeling, 11, 65, 71
repair, 2, 14, 28, 43, 70, 86, 112, 121, 124, 130, 131, 134
reparation, 33, 107
repetitions, 3
research and development, vii
residues, 3, 58
resilience, 49
resin, 35, 111
resistance, 11, 18, 39, 40, 47, 49, 50, 89, 148
restorative dentistry, 78
restorative materials, 108
retention, 65
Reynolds, 135
rhenium, 86
rheological properties, 47
rheology, 61
rigidity, 13, 21, 35
risk, 21
risks, 1
rods, 67, 122, 151
rolling, 20
room temperature, 29, 46, 102, 121, 128

roughness, 37

S

safety, 136
saline, 27, 67
salt, 29
salts, 63, 109, 150
sample, 29
sand, 34
scaffold, 2, 16, 24, 25, 51, 67, 78, 81, 83, 93, 94, 98, 99, 102, 106, 107, 112, 114, 116, 118, 121, 127, 130, 131, 132, 136
seed, 103
seeding, 76
Self, 77, 115
self-assembly, 149, 151
self-organization, 41, 58, 104, 125, 126
sensing, 55
sensitivity, 63
separation, 23, 31, 38, 115
serum, 3, 22, 45, 131
serum albumin, 45, 131
services, iv
shape, 9, 30, 45, 48
shaping, 21, 27, 46
shear, 3
sheep, 32, 134
shock, 140
short period, 37
short-range, 3
short-term, 87
sign, 58
silane, 62, 110, 150
silica, 48, 49, 142
silicon, 43, 48, 62, 115, 134, 137
silk, 16, 46, 93, 120, 121, 134
silver, 51, 143
similarity, 11, 41
simulated body fluid, 25, 31, 43, 93, 131, 144, 145
simulation, 42, 127
Singapore, 76
singular, 40

sintering, 24, 49, 51, 52, 64, 127, 136, 137, 140, 143, 145, 146, 148
sites, 1, 4, 25, 41, 42, 48, 58, 61, 63, 75, 126
sodium, 4, 35, 64
sol-gel, 98, 119, 136, 141, 142, 149
solid phase, 46
solidification, 9
solubility, 13, 31, 37
solvent, 22, 23, 26, 28, 31, 43, 63
solvents, 26
sorption, 108
soy, 16
spacers, 18
spatial, 39
species, 1, 4
spectroscopy, 33, 59, 60, 61, 64
spheres, 46, 60, 94, 103
spinal fusion, 124
spine, 67, 105
sponges, 43, 105
stability, 4, 8, 9, 12, 29, 38, 39, 67, 111, 119, 136, 138, 140, 146, 148
stabilize, 18, 35
stages, 20, 46, 50, 66
stainless steel, 16
stainless steels, 16
standards, 70
starch, 16, 28, 67, 100, 101, 152
sterilization, 13, 70, 82
stiffness, 3, 4, 8, 24, 27, 28, 65, 91
strain, 3, 4, 53
strategies, 23, 50
stress, 3, 4, 15, 29, 53, 57, 65
stress-strain curves, 53
stretching, 59
stromal, 136
stromal cells, 136
strontium, 109
structuring, 104
styrene, 46, 97
substances, 4, 7, 50
substitutes, viii, 1, 4, 19, 28, 42, 46, 47, 76, 80, 92, 107, 112
substitution, 1, 4, 23, 28, 33, 37, 43, 75, 76, 88, 99, 102, 135

Index

substrates, 53, 79, 92, 125, 126, 149
sugar, 94
sulfate, 48, 92, 145
sulphate, 125, 145
Sun, 80, 94, 103, 105, 107, 112, 113, 114, 121, 142
supercritical, 24, 94
supercritical carbon dioxide, 94
superiority, 40
supramolecular, 45, 132
surface area, 21, 37, 46
surface energy, 38
surface layer, 33
surface modification, 33, 97
surface properties, 37, 70
surface roughness, 37
surface treatment, 88
surfactant, 26, 102
surgeons, 1
surgery, 14, 40, 43, 69, 71, 73, 82, 84, 90, 105, 133, 135, 152
Surgery, 84
surgical, 2, 20, 31, 47, 75, 83, 84, 141
survival, 23
suspensions, 47
swelling, 39
synthesis, 11, 16, 41, 79, 86, 93, 98, 102, 108, 113, 123, 126, 149
synthetic bone, 2, 123
synthetic polymers, 14, 16, 85, 120

T

tantalum, 86
teeth, 2, 11, 44
temperature, 23, 46, 63, 64, 99, 106, 112, 149
tendon, 124
tensile, 2, 16, 21, 25, 34, 45, 87, 97, 110
tensile strength, 2, 16, 21, 25, 34, 45
tension, 3, 97
tetracycline, 29, 44
thawing, 23
therapy, 130
thermal stability, 136
thermal treatment, 48
thermally induced phase separation, 23, 38, 115
thermogravimetric, 59
thermogravimetric analysis, 59
thermoplastic, 101
thermoplastics, 100
thin film, 141
thin films, 141
Thomson, 78, 81
three-dimensional, 4, 76, 101, 116, 130, 135
time consuming, 69
Tissue Engineering, 77, 81, 152
tissue-engineering, 85
titanates, 62, 64
titania, 48, 98, 136, 137, 140, 141, 142, 149
titanium, 44, 52, 54, 86, 92, 98, 119, 125, 128, 131, 140, 141, 142, 144, 146, 148, 149
Titanium, 16
titanium dioxide, 141
TJ, 98, 113, 122, 127, 141, 147
TM, 105, 145
Tokyo, 81
torque, 29
toughness, 2, 3, 16, 17, 18, 21, 49, 51, 77, 142
toxic, 1, 11, 13, 14, 26
toxic products, 13
toxicity, 4, 32, 65, 84, 106
trabecular bone, 14, 28, 52, 78
trade, 20
transducer, 55
transfer, 2, 40, 57, 61, 65
transformation, 18, 63, 139
transition, 53
transition temperature, 16
transitions, 13
transmission, 42, 64
transmission electron microscopy, 64
transmits, 57
transparent, 13, 24
transplantation, 1
transport, 8
trauma, 4, 75, 84

trial, 124, 127, 136
tribological, 18
trochlear, 43
trypsin, 67
tumor, 1
tumor cells, 1
two-dimensional, 17
type 1 collagen, 126
tyrosine, 14

U

ultrastructure, 101
uniform, 8, 20, 41, 42, 45, 54

V

vacuum, 26
valence, 9
values, 12, 25, 50, 129
van der Waals, 57
van der Waals forces, 57
vancomycin, 32
variables, 69
vascularization, 11, 71
vector, 132
velocity, 141
vertebrae, 18
vibration, 137
vinylchloride, 13
viral infection, 1
viscoelastic properties, 13
viscosity, 23, 47, 63

W

water, 3, 25, 28, 33, 34, 36, 43, 45, 47, 50, 62, 86, 108, 121, 142
water absorption, 25
water sorption, 108
water-soluble, 36, 45, 47
water-soluble polymers, 47
weakness, 17
wear, 18, 50, 82, 144
web, 40
weight loss, 67
wettability, 23
wetting, 37
wires, 87
workability, 35
workers, 45

X

xenograft, 125
xenografts, 1
X-ray diffraction, 60, 61

Y

yield, 22, 28, 29
yttria-stabilized zirconia, 138, 139
yttrium, 87

Z

zirconia, 18, 48, 51, 53, 64, 80, 87, 136, 137, 138, 139, 140, 145, 148, 151, 153